Decio Giulio Riccardo Carugati

Pininfarina

Identity of a Design

Electa

Translated by
Giacomo Donis, Venice

In the 1950s the Museum of Modern Art in New York placed the Pininfarina Cisitalia in its permanent collection. For the first time, as "sculpture in motion," according to Arthur Drexler's definition, a car had been recognized as an object of art.

Since then the world has been changing more rapidly than ever before. The accelerated evolution of science and technology, including informatics, has led to profound modifications in the cultural fabric and in customs.

Nevertheless, it cannot be denied that certain historical prejudices still survive today, related to the stylistic models of medieval Scholasticism and its distinction between greater and lesser arts. On the basis of this presumed greater or lesser "nobility" of the arts, there is a tendency to deny in fact the status of "artist" and person of culture to the industrial designer and, all the more so, to the designer of such a popular consumer object as the car is today.

Why is there such great distrust and intellectual hostility with regard to the very product that more than any other has given rise to our modern industrial culture? What, indeed, is more complex in design than the car, bound up as it is with the organization of labor, with the question of mass production, with the control of costs—in short, with the full range of conditions that constitute what we mean by "industrial design"?

What accounts for this sort of "repression" of car design on the part of most design historians and scholars?

It has been observed how the culture of design has split in two, with the intellectual stratum of designers, graphic artists and architects on one side, and the direct development of industrial productive culture on the other.

I quote from the memoirs of my grandfather Pinin Farina: "I'd gotten to a point in my work where technical know-how and pure form didn't count anymore; what counted was the ability to keep the two of them together, effortlessly or at least without showing the effort. The Cisitalia, when it came out, in the autumn of 1947, was praised for its design, but even this seems to me a misunderstanding; the coachbuilder is above all a builder, but if he pays too much attention to the cosmetics of his product he'll forget its function… On the other side there were the technicians, who considered my work to be a simple fact of tailoring. I rejoined: 'If it were nothing more than a technical fact, the car would be ugly,' and I keep repeating this today because there's still a need to say it."

Around thirty years later my father, on the occasion of the Compasso d'Oro career prize that ADI awarded him in 1994, put it this way:

"I've said more than once that the designer can be considered the artist of our times, the bearer of a complex of cultural values that are at the same time formal and technical. It is in fact clear that industrial design is not simply a formal and figurative expression suspended in the void of pure aesthetics; on the contrary, it is involved both with scientific and technological innovation and with the constraints of mass production."

The Italian word "progetto" corresponds to what in English is called "design." Oli and Devoto, co-authors of the Italian dictionary that bears their name, define the word "progetto" as "a plan, generally accompanied by a study relative to the possibility of its realization and execution." The "project" is thus a source of simplification, and its identity manifests itself over time as that form of complexity which is the most opportune, possible and feasible. So, in accepting the Publisher's invitation to write this book, I have undertaken to explore the meaning of the identity of Pininfarina design and of the qualities that distinguish it. I have done so by telling the story of the bodywork of a number of extraordinary cars. Taking these examples as way stations of a process of technical and formal evolution, I have then attempted to let the protagonists of my story speak for themselves. Thus Sergio tells us about his father, Pinin, and the two of them, along with others, tell us about Pininfarina design, placing it within the social, political, cultural and artistic climate in which it was born and through which it evolved.

In the 1950s when, for example, Arteluce (founded by the designer-entrepreneur Gino Sarfatti) was already displaying the unmistakable identity of its design in its limited series of lamps, Pininfarina (founded by the designer-entrepreneur Battista Farina) was showing its own identity, mass-producing niche automobiles that were not a whit less unmistakable. It is not by chance that works by both designers are on display, as archetypes, in New York's Museum of Modern Art. As Gillo Dorfles noted in 1958, "It was not before the 1930s that anyone sought to individuate that particular category of design for industry (i.e., for objects that were mass-produced through industrialized methods and systems) where the technical datum is combined from the outset with an aesthetic element." But Pininfarina precedes Arteluce when, right back in the 1930s, it began to give voice to the same declaration of identity it still maintains today, producing its custom-built cars, its mass-produced cars, and its research prototypes. It would be misleading, however, to treat these categories as pigeon-holes. A remark of Lucio Fontana's comes to mind: "A butterfly in space excites my imagination." Apparently simple, the light form of the butterfly reveals the complexity of design. I look at a photo of the Dino Berlinetta Speciale of 1965 and I realize that even today, the 360 Modena fires my imagination in the same way, especially if I compare it with the normalized and normalizing forms now dotting the automotive landscape, accompanied by the rhetoric that justifies their insipid invasion.

A great sense of lightness: this is what distinguishes the identity of Pininfarina design in the forms that reveal its complexity. And this sense comes across in the sequence of images, assiduously arranged by Stefano Tosi, that Francesco Pagni, in charge of the historical and documentary archives of Pininfarina, has skillfully matched to the theme of my narration. I thank both of them for their collaboration and, through Stefano Peccatori, I thank the entire editorial staff of Electa.

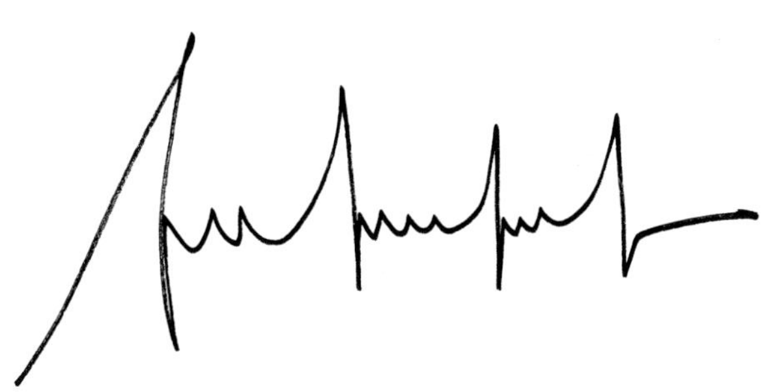

Contents

In this volume text and images follow two different developments: the automobiles described in the text are illustrated in the sequence that is indicated in the table of contents; the running titles in the lower part of each page, together with the title of the chapter, state the name of the model illustrated.

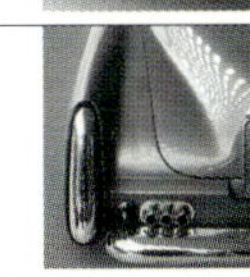

Custom-Built Cars

Detail of the Cadillac V16 (1931) restored in Reno (Nevada) by the collector Bob Lee.

"Asia," decorative panel made for the Colonial Exhibition in Paris, 1931, showing a sumptuous train of elephants in India.

On November 30, 1929, incredulous and stunned, the magazine *L'auto italiana* reported the news of the latest General Motors production: "It would appear that Cadillac is presenting, on various markets, a model with a V16-cylinder engine. This new car has been designed and fine tuned by the engineer who created the 8-cylinder Cadillac and Salle." The conclusion of the report is ironic, and even the syntax is slightly out of control: "The management of the Cadillac Motor Co. of Detroit thinks there exists a large public likely to take an interest in cars with exceptional characteristics, with no regard for the price." But the roaring 1920s, the jazz age, the illusion of an endless party had all collapsed in a single day, the 25th of October, Black Friday, with the Stock Market crash.

The age, however, had had its poet: Francis Scott Fitzgerald (1896–1940) set out to immortalize its myth in 1925 with his novel *The Great Gatsby*. "There was music from my neighbor's house through the summer nights. In his blue gardens men and girls came and went like moths among the whisperings and the champagne and the stars. At high tide in the afternoon I watched his guests diving from the tower of his raft, or taking the sun on the hot

sand of his beach while his two motor-boats slit the waters of the Sound, drawing aquaplanes over cataracts of foam... At least once a fortnight a corps of caterers came down with several hundred feet of canvas and enough colored lights to make a Christmas tree of Gatsby's enormous garden. On buffet tables, garnished with glistening hors-d'œuvres, spiced baked hams crowded against salads of harlequin designs and pastry pigs and turkeys bewitched to a dark gold. In the main hall a bar with a real brass rail was set up, and stocked with gins and liquors and with cordials so long forgotten that most of the female guests were too young to know one from another." The young stockbroker Nick Carraway is renting a house on Long Island, not far from the mansion of Jay Gatsby, whose acquaintance he will soon have occasion to make.

The streamlined body stands out above the demarcation line of the Cadillac V16.

Cadillac V16 (Maharajah of Orchha), 1931

It was unquestionably one of the most fascinating cars my father designed in that period, or in any period for that matter.

Sergio Pininfarina

Picture of the Victoria from an advertisement of the Pinin Farina coachbuilder, 1933.

A typical image of the 1930s: a charming young lady with a greyhound at her side.

"At nine o'clock, one morning late in July, Gatsby's gorgeous car lurched up the rocky drive to my door and gave out a burst of melody from its three-noted horn. It was the first time he had called on me, though I had gone to two of his parties, mounted in his hydroplane, and, at his urgent invitation, made frequent use of his beach. 'Good morning, old sport. You're having lunch with me to-day and I thought we'd ride up together.' He was balancing himself on the dashboard of his car with that resourcefulness of movement that is so peculiarly American—that comes, I suppose, with the absence of lifting work in youth and, even more, with the formless grace of our nervous, sporadic games. This quality was continually breaking through his punctilious manner in the shape of restlessness. He was never quite still; there was always a tapping foot somewhere or the impatient opening and closing of a hand.

He saw me looking with admiration at his car. 'It's pretty, isn't it, old sport?' He jumped off to give me a better view. 'Haven't you ever seen it before?' I'd seen it. Everybody had seen it. It was a rich cream color, bright with nickel, swollen here and there in its monstrous length with triumphant hat-boxes and supper-boxes and tool-boxes, and terraced with a labyrinth of wind-shields that mirrored a dozen suns. Sitting down behind many layers of glass in a sort of green leather conservatory, we started to town."

The narrative crescendo explodes in the definition of the length of the car: "monstrous," it clearly implies the extraordinary power of the engine hidden under its hood. The chrome accentuates the overall effect, exaggerating, amplifying the object's visual impact. The wealth of boxes, the endless windshields and the voluptuous interior complete the picture.

Fitzgerald was to return to the themes of the roaring 1920s in *Tender is the Night*, the story of a group of American expatriates in France. "The party that night moved with the speed of a slapstick comedy. They were twelve, they were sixteen, they were quartets in separate motors bound on a quick odyssey over Paris. Everything had been foreseen. People joined them as if by magic, accompanied them as specialists, almost guides, through a phase of the evening, dropped out and were succeeded by other people, so that it appeared as if the freshness of each one had been husbanded for them all day. Rosemary appreciated how different it was from any party in Hollywood, no matter how splendid in scale. There was, among many diversions, the car of the Shah of Persia. Where Dick had commandeered this vehicle, what bribery was employed, these were facts of irrelevance. Rosemary accepted it as

Alfa Romeo Coupé Victoria, 1933

The Victoria's sharply angled windshield projects the interior towards the rear.

merely a new facet of the fabulous, which for two years had filled her life. The car had been built on a special chassis in America. Its wheels were of silver, so was the radiator. The inside of the body was inlaid with innumerable brilliants, which would be replaced with true gems by the court jeweler when the car arrived in Teheran the following week. There was only one real seat in the back, because the Shah must ride alone, so they took turns riding in it and sitting on the marten fur that covered the floor."

Fitzgerald began writing the novel in 1932. The previous year an Indian Maharajah made up his mind to add an American car to his collection of automobiles—but designed by an Italian coachbuilder! "Who does not know the Cavalier Battista Farina, the superb author of so many masterpieces that have helped gain international renown for the Italian art of beautiful bodywork? Who does not know that the good taste and skill of this exceptional coachbuilder have satisfied on countless occasions the taste, the demands and the desires of the most illustrious nobles of the old world, of blue-blooded Princes?" This was how *L'auto italiana*, in June 1930, saluted the birth of the factory in Corso Trapani in Turin that bore the name "Pinin," the nickname by which the Cavalier Battista Farina was commonly known. "Modernly and grandiosely equipped for the production of highly luxurious car-bodies. Pinin Farina will certainly be treating us with new prodigies of his art. This, moreover, is his intention; and this is the wish with which our magazine salutes the beginning of his renewed form of activity."

If the Indian Prince was looking for an Italian coachbuilder, Pinin had everything it took to fit the bill. What's more, as *L'auto italiana* of 31 January 1930 declares, at the New York Motor Show the V16 engine had become an indisputable reality. "The V-angled (45-degrees, naturally) 16-cylinder engine proves to be quite compact despite the arrangement of the carburetors and of the pipes outside the V, with a bore of about 76 cm and a stroke of 100 cm, giving around 180 hp at 3200 rpm; this enormous power cannot help but guarantee the exceptional qualities of the car. The Cadillac V16 is unquestionably one of the most interesting cars on display at the Gran Central Palace."

But who in the world was that Prince, sovereign of a region of central India called Madhya Pradesh, Lord of Orchha? "74 miles from Gwalior and 10 miles from Jbansi, formerly the capital of the Bundela rajputs (a Hindu military caste), this medieval town on the banks of the Betwa is today a small village with a sleepy air, which does however possess many architectonic gems. Most noteworthy is the Jehangir Mahal, an attractive building with crenellated pavilions, decorated at various points with blue and yellow ceramic tiles, and with an enormous entrance-gate with elephants on either side. The ceiling and the walls of the verandah that encircles the four-

The line of the front fender is extremely pure, stretching out until it takes in the running-board and then rising light and soft towards the rear.

Alfa Romeo Coupé Victoria, 1933

The central body of the car remains narrow, for its full length, and is clearly reminiscent of an airplane wing.

The headlights of the Pescara, incorporated in and protected by grills, repeat the design of the radiator grill.

The two images opposite page are dedicated to the aeronautics of the 1930s: cover of the magazine "L'ala d'Italia," 1932 and postcard for the 10th Anniversary of the Air Cruise Rome Chicago New York (Italo Balbo), 1933.

story central tower of the Laxminaryan temple have a great abundance of murals painted in browns, ochres blacks and greens, depicting battles both mythical and real (taking place in 1857), as well as abstract decorations and portraits." This is what the Madhya Pradesh State Tourism Development Corporation has to tell us today. What was the rank of the Maharajah of Orchha, perhaps the last of the Mughals, Bahadur-Shah? How many cannon shots, how many salvos were fired to salute his official appearances? "The sovereign vassals of India are entitled to a maximum of twenty-one cannon shots, a privilege that is enjoyed by only five Maharajahs: of Baroda, of Gwalior, of Mysore, of Jammu and Kashemir (which form a single state), and of Hyderabad. Six other princes have a right to nineteen cannon shots, thirteen to seventeen salvos, seventeen to fifteen shots, sixteen to thirteen and, finally, nineteen minor princelings are entitled to a mere eleven shots. None of the other Indian Princes can aspire to any artillery salute whatsoever, and there are over six-hundred of them, including His Highness the Aga Khan and the Maharajah of Burdwan, who are not ruling heads of state. The Aga Khan, for example, is simply a spiritual leader… It is clear, then, that England wished to create for itself in India a sort of feudal empire, which, alas, did not satisfy anyone—neither the Indian princes, nor the natives who were directly subject to English rule."

So, while P.G. Jansen, in *La Sera*, 14 April 1939, goes by the book in condemning Britain's Indian policy, which exploited even the most backward of situations to keep hold of its dominions, in Paris ten years earlier Cartier was resetting, to his taste, the jewels of the Maharajah of Pantiala's crown, and the public was dreaming Oriental dreams. We find this dream even in the lyrics of the popular songs collected in the edition of *Fonografo Italiano 1890–1940*. "Bengala," for example, goes: "Down there / the fragile virtue / is given like a flower./ The heart / of the malicious Hindu / burns like a Bengal fire./ Every snake charmer knows / under the spell you will fall in love / and you will have to stay down there / under the sky of the Bengal fire./ When the dawn purples with gold / without mandolas a thousand Indianolas / in the sugarcane / sing sweet words to the sun / and salute love as it departs / on the elephant of the Rajah." A fairy-tale climate, then, right out of an operetta. Anything was possible.

"But it's hard to imagine that the Maharajah of Orchha had ever heard of my father," Sergio remarks. "Perhaps the General Motors agent in Rome, or in Milan, or even the Indian Embassy in Italy, had taken on the job of

Alfa Romeo Pescara, 1935

preparing the car and then commissioned my father for it. One thing is sure: in the future of Pininfarina there was, of all things, a Cadillac! Otherwise, why in the world would my father, to produce a one-of-a-kind custom-built car, have turned his back on all the Italian chassis—Lancia, Alfa Romeo, Isotta Fraschini—that he had often used before, and with excellent results? Sure, he figured that a big luxury American chassis couldn't help but boost the reputation of his brand-new factory. But this had nothing to do with deferring to the powers that be; rather, from the very beginning he wanted to declare his love for the automobile, and especially for its aerodynamic dimension. If, in fact, the Indian prince had to go tiger hunting with that very car—well, no problem, he'd fit it out with special compartments for the shotguns. The important thing, however, was that the object should absolutely be the fruit of his own research, and not just a replacement for that canopy on the elephant's back which had traditionally served the purpose. The Maharajah wanted a Cadillac? Well that's what he would get—but as my father understood it. So General Motors delivered to Pininfarina a Cadillac V16 chassis, complete with radiator grill (the size of the louvers matching the mastodonic radiator), wheels and engine. My father loved low, sleek lines and felt the limit of that structure so high above the ground that, especially in the front, prevented him from asserting his own criteria. He decided, ideally, to cut the car in two, creating a top and a bottom half. Going to work on the bottom half, he eliminated the running-boards and wrapped the side members, while lightening the structural impact with a series of false air intakes. As a result, on the top half the body now looks streamlined—or as streamlined as the radiator grill will allow. And this solution was something absolutely new, since American design at that time would unquestionably have preferred the single section, never one body on another, resulting in a general heaviness of the line. My father also knew that, in an open car, the smaller the doors, the more solid and stable the body. So he downplayed the front doors, which with the streamlined tail call to mind two-seater sports cars, and the rear doors as well, which hid two rear seats, to be unveiled when required. The front seats were for the driver and the escort; in the back sat the Maharajah who, when he caught sight of the tiger could stand up, take his shotguns from their special compartments, take aim and fire! The car, however, gives proof of a research into design that goes far beyond its utilitarian purpose. Completed in four months, painted an off-white (partly brushed on by hand

The Lancia Aprilia Aerodinamica, on the race track and on display at the 1937 Milan Motor Show, features details of the body of great stylistic finesse.

and partly sprayed with the sort of pistol in use at the time), the car was shipped off to India. As the story goes, during the war, afraid that the English would confiscate it, the Maharajah of Orchha had the car buried under a lot of straw and earth somewhere in the countryside. After the war and, presumably, after the death of the Maharajah, the car was sold to an American, and very cheaply at that, who, in turn, sold it to Bob Lee, a car-collector from Reno. Bob owns more than five hundred cars of various periods, perfectly restored and in working order. American-style restoration work, though, is completely different from the way it's done in Europe, which is more refined, more responsible. In Europe, an auto is recuperated, brought back to life, with respect for its original features and the materials with which it was built, especially as regards the car's interior. In America they're mostly interested in luxury, in gratuitous chrome and precious leather. They end up with a car that may be stupendous, but that has little to do with what the car was really like in its heyday."

A funny story: the Cadillac V16 was born at the tail-end of the roaring 1920s, "in its monstrous length" like the car of Fitzgerald's Gatsby, and perhaps to satisfy a whim of the Shah of Persia, as the crowning jewel of his collection of automotive gems. But, none the less, it brought to India, to the Maharajah of Orchha, the innovatory criteria of an Italian who, in those early years, already understood the car as *design*; and that very car, despite the incautious and unrefined restoration it went through, still bears witness today to his brilliant intuition. It was not Pinin, in fact, who insisted that his first cars should be custom-built, but his clients who demanded it, ruling out any possibility of mass-produced cars. For his clients in those days ordered one single auto, not ten, a hundred, a thousand, as the automotive industries were to do later on. "My father," Sergio recalls, "was convinced—and it's still true today—that you obtain quality through precision craftsmanship, but even better quality through mass production." In his essay, "Protodesign italiano," Anty Pansera explains: "If the unique object—the art object— was a characteristic feature of the 1920s, when the question of mass production was just being broached, on the sly, in certain sectors, the 1930s witnessed the gradual but clear rise of what we now call design, as it was first introduced at the *Mostra Internazionale della Produzione in serie* [International Exhibition of Mass Production] at the 7th Milan Triennale. But the 'design' of the 1930s did not fit today's definition of the term because it was not yet industrial design, but rather a sort of mechanized craftsmanship that had established itself in the Milan area and northern Italy generally. So the proper word for it would be 'protodesign.' The unique object, on the other hand, not only continued to have intrinsic value; it also had an influence on mass production. This was true as regards 'small objects' (of which there were many in the editions of the Triennale), whose history is close, or linked, to that of 'applied arts,' and whose quarrel with 'industry,' in Italy too, had been longer lasting than was the case with other 'products,' which thus lent themselves more readily to mass production. And it was also true of larger objects, such as furniture and cars."

In April of 1933, in Germany, the School of Weimar came, definitively, to an end: the National Socialist-dominated City Council of Dessau, where the Bauhaus had moved back in 1927, ordered the closing of its seat. Klaus Stefan Leuschel, in *Design. Developments in Germany*,

In 1936 Pinin Farina, at the wheel of his Aprilia Areodinamica, hits 100 mph on the road from Rome to Ostia.

The original mock-up of the Aprilia Aerodinamica.

writes: "In the initial phase we already find, along with Walter Gropius, the painter Lyonel Feininger and the artists Johannes Itten and Gerhardt Marcks among others, who were later followed by Paul Klee and Wassily Kandinsky, the Hungarian artist Laszlo Moholy-Nagy, Oskar Schlemmer and the Swiss architect Hannes Meyer. In the last period of the existence of the Bauhaus, such architects as Ludwig Hilberseimer, Ludwig Mies van der Rohe and the Dutchman Mart Stam taught there. All these personalities were a guarantee that the production of the Bauhaus should cover the fullest possible range of artistic expression... With craftsmanship as its base, the activities of the school were marked by a profound reflection upon the methods of industrial manufacture. Even if the production of the Bauhaus—in large measure due to a lack of interest on the part of industry—was never more than very small-scale, it was nonetheless an expression of the industrial age that established an image, which is still valid today, of an everyday culture set in the spirit of its time... In Walter Gropius this ideal of an everyday culture in tune with its time manifests itself in the impulse of art to move towards industry and, accordingly, in the wish to integrate art into life. The moving force was and remained the attempt to solve, or at least to harmonize, social problems by means of aesthetic production... The Bauhaus in fact had demanded from the start that products be justified by their social meaning and function, while contributing to a form of the environment that was more respectful of human dignity."

The teachings of the Bauhaus were truly ahead of their time, especially back in its first years, in the 1920s (Gropius founded the Staatlisches Bauhaus Weimar in 1919). "It was not before the 1930s that anyone sought to individuate that particular category of design for industry (i.e., for objects to be mass-produced by means of industrialized methods and systems) where the technical datum is combined from the outset with an aesthetic element. And this statement by Gillo Dorfles," as Anty Pansera remarks, "from 1958 but still extremely topical, throws into relief the second bastion of interpretation and of choice: namely, the will of the designer that the design he creates should contain within itself (perhaps in a latent state, but one that is eminently realizable) that quality of uniqueness and of artistic individuality which will distinguish it from every other design and which constitutes its real identity."

In his cars of the 1930s, while tied down to his clients demand for custom-design, Pinin was already expressing his own will as a designer, concretely realizing that artistic uniqueness which

The imposingly large convertible top of the Lancia Astura Bocca is opened by a hydraulic servo-assisted mechanism, patented by Farina.

will prove to be the identity of his product. This identity had already been foretold by Pinin himself when, in the spring of 1920, visiting the United States, and Detroit in particular, in coming into contact with the reality of America he commented: America is certainly the breeding ground of the automobile, but the questions of form and line, which make a car beautiful in its usefulness, find in Italy, despite it all, a more suitable terrain for inventive solutions. "I go to the Ford works, which had already moved almost entirely to River Rouge, and I join up with fifty-odd visitors." Pinin is telling his story to Ernesto Caballo, for the book *Nato con l'automobile* [Born with the Automobile]. "So I follow the conveyors from one department to another as they wend their way steadily with materials and various parts; the workers take what they need from them, even the chassis move through the various stages of assembly. This scene alone was worth the entire cost of my trip to America. A guide was explaining to me in French that Ford had gotten the idea of the conveyors when he visited the big slaughter-houses of Chicago; after it was killed the animal was hung up on a moving cable, one person removes the hair, another one takes out the entrails, someone else cuts it up, so on and so forth… I didn't follow the details but rather the rhythm of the work, and everything went smoothly. There were conveyors of various forms and for various purposes, and every assembly line was set up functionally. The whole thing was made easier by the fact that they were only making one product, the 'T' Model. It was astonishing."

Pinin is fascinated by the system of mass production that he sees in America in 1920; but he also feels that, at least from a formal point of view, the excellence of the system is not matched by that of the product. And he is convinced that, as the demand for cars in Italy and the rest of Europe increases, there will be an even greater need for designers capable of reconciling the dual concept of technology and aesthetics—of giving form to function. He already understands, in 1920, the need to move from a culture of the unique object, *l'objet d'art*, to the culture of an object that repeats its artistic uniqueness. He also knows, however, that to accomplish this the first thing that is needed is a clear design, and that all the means of its realization come later. Here, then, we have the reason why he criticized the American system of production that had so astonished him, in the years that led up to the founding of the company that will bear his name.

But what were the social, cultural and artistic tensions that shaped the fundamental intuitions of Pinin Farina? In his essay "L'arte

Lancia Astura Bocca, 1937

The car is imposing due to its extraordinary size, over 16 feet long, which is unmitigated by attempts to hide the dimensions but taken as a smooth monolithic mass.

Lorenzo Ramaciotti

all'ordine del giorno" [Art on the Agenda], Vittorio Fagone writes: "We can look at the 1930s as the years of the spread of the eclectic 'Novecento movement,' as the heroic age of abstractionism in Milan and Como, as the period of an ambitious attempt to express a social and political art through mural painting. In the same period Futurism continued the work of spreading its message and experimenting with new forms, while a generation of younger artists, reassembling the terms of a romantic poetics, looked for new ways to oppose the Novecento group. The 1930s were also the years of rational architecture, of the foundation of new cities, episodes that went hand-in-hand with the practice of architectural monumentalism. And the 1930s were also the years in which statutes were laid down for the major exhibitions (the Venice Biennale, the Milan Triennale, and the new Rome Quadriennale), and a vast network of support and assistance for the visual arts was drawn up: for example, the expositions of the *Sindacato Fascista degli Artisti* [Fascist Union of Artists], the *Ufficio per l'Arte Contemporanea* [Office for Contemporary Art], the 2 per cent law. The interest of the general public in art was stimulated by major exhibitions of new architects and artists of the various schools. The strengthening of political mass-organizations (the *Gioventù Italiana del Littorio* [Fascist Youth Organization], the *Opera Nazionale del Dopolavoro* [National Institute of Workers' Recreational Clubs]) and the astute use of new media (the radio, the cinema) are elements that can be linked to the policy of patronage of and incentives for the arts."

The Lancia Astura Bocca is a super-luxury car, closer to the pampered world of "white telephones" typical of the rich bourgeoisie that people saw in the movies.

Augusto Morello, in his monograph *Sergio Pininfarina*, recounts the artistic, social and cultural atmosphere of the 1930s in the city of Turin. "Turin was going through years of great cultural vitality (the visual arts, architecture and city-planning, literature and philosophy, political, economical and social theories, physics and chemistry); but its entrepreneurs were not inclined to bring together the two cultures (the humanistic and the technical) of Charles P. Snow, as had instead been the case in Milan at least until the 1980s. And so, since the glorious Academy of the Sciences of Lagrange, of Cauchy, of the corresponding member Gauss, of Avogadro and of Peano, in fact never communicated with the Albertine Academy of Fine Arts, art in Turin had little or nothing to do with industry, except for the usual patronage or promotional activities. Yes, Turin did take in solitary artists from all over, such as Luigi Spazzapan from Gorizia (whose influences on the designer Ettore Sotssas Jr. are well known);

HUILE

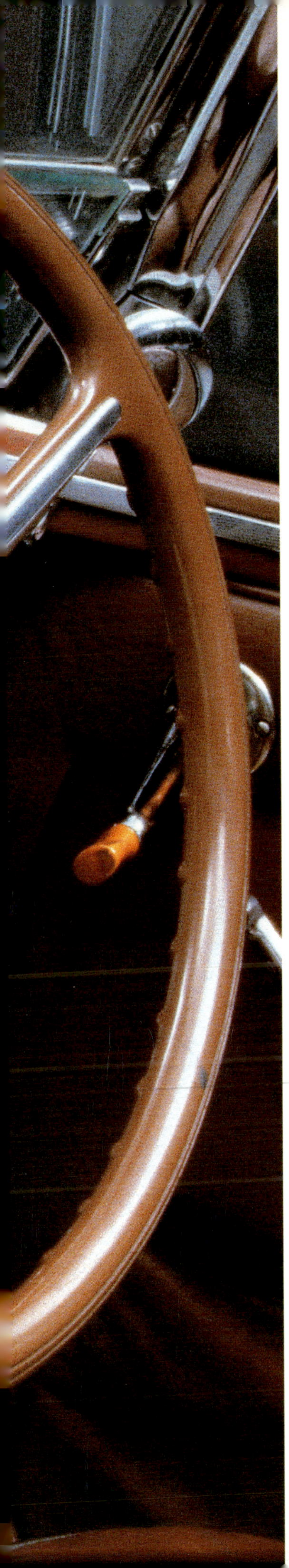

A detail of the refined dashboard of the Lancia Astura Bocca, designed in 1937.

the famous Group of Six was formed there (Luigi Menzio from Sardinia, Nicola Galante from the Abruzzi, Enrico Paolucci from Liguria, Jessie Boswell from England, plus Gigi Chessa and Carlo Levi); and in Turin in 1928 Felice Casorati, from Novara, opened the first exhibition of abstract art in Italy, and began to exercise his long-lasting influence on the city's art scene. Then, there were art patrons such as Riccardo Gualino who, before his internment by the Fascists, commissioned Giuseppe Pagano (with Gino Levi Montalcini) to design works of architecture, brought Lionello Venturi back to Turin (and Venturi influenced the young Turin-native Argan, as well as the Neapolitan Edoardo Persico in his short stay in Turin), and invited innovators in music and prose to his misunderstood Theater. And there was the women's group for the arts, the *Pro Cultura Femminile*, that called in Arnold Schonberg in 1924 to conduct *Pierrot Lunaire*, and the architect and designer Carlo Mollino who was just beginning his isolated and untimely investigations... Outside the world of the automobile, the only Piedmontese of international note who had sensed a need for the effective synthesis of the two cultures that went beyond mere patronage was, after the war, Adriano Olivetti; but, from his headquarters in Ivrea, he in fact gravitated towards Milan, thus bypassing Turin... It was in Turin, however, that—with and without Fiat—we find much of that invasive cultural adventure which managed to turn a number one commodity like the car into an occasion for interdisciplinary evolution."

Filippo Tommaso Marinetti had been ahead of his time in envisioning the invasion of the new automobile culture which, however, for him was linked to the myth of *speed*. On February 20, 1909, publishing his "Futurist Manifesto" in the *Le Figaro*, he wrote: "We affirm that the magnificence of the world has been enriched by a new beauty: the beauty of speed. A racing car with its hood adorned with large pipes like fire-breathing snakes... a roaring car, which seems to run to machine-gun fire, is more beautiful than the Victory of Samothrace. We want to praise the man who holds the steering wheel, whose ideal shaft runs right through the Earth, which is itself racing on the track of its orbit." So here the moving object has the form of its engine, launched on an endless quest for the peaks of velocity. In *Il Futurismo e l'automobile* Rossana Bossaglia writes: "It is not by chance that Marinetti in his Manifesto of 1909 compares the automobile with the Victory of Samothrace. In that moment he says two things: first, that art of the past is nothing compared to the art that can be projected into the future; second, that no work of art can compete with a technological work that contains the new artisticity within it. In any case for his comparison he has chosen an extremely dynamic model. In fact he takes the Nike of Samothrace, not the Mona Lisa or the divinities of the Parthenon; he takes one of the most beautiful dynamical sculptures in the world. The Nike of Samothrace that comes right at us as we go up the staircase in the Louvre is really impressive. Marinetti doesn't say the Nike is ugly, he says that the automobile is more beautiful even than the Victory of Samothrace."

Marinetti, however, in 1909 was glorifying a speed that was far out of reach of the automobile of the day. But the true meaning of his "invasive cultural adventure" is to be grasped, as Augusto Morello makes clear, "in the first rational (in a manner of speaking) experiments, whose aerodynamical stages go back to Jenatzy,

Renato Rascel and Sergio Pininfarina discuss details of the delivery of the PF 200.

the Belgian, with his tear-drop-shaped six-seater, which Marco Ricotti in Milan commissioned in 1913 from the coachbuilder Castagna on an Alfa Romeo, and above all the Bizzarra of Michele Lanza in 1925." That hood adorned with large pipes like fire-breathing snakes, a romantic personification of speed, could not be souped up *ad infinitum* in search of thrills. The self-propelled object had to work together as a whole if it was to reach the highest peaks. And the only way to attain this was by studying the shape most fit to express the meaning of speed. This was Pinin's conviction, and it led him to design his bodies according to aerodynamic intuitions that he experimented empirically.

Pinin, then, wasn't looking to his cars for a gleaming affirmation of speed. Rather, he was seeking in nature the confirmation of a possibility: shapes as streamlined as fish and as agile as greyhounds. So here is the meaning of the Alfa Romeo Coupé Victoria, the 8-cylinder 2300 of 1933: "It was unquestionably one of the most fascinating cars my father designed in that period, or in any period for that matter," Sergio remarks. "You note right away the same quest for sleekness as in the Cadillac V16 for the Maharajah of Orchha, despite the constraints of that case, and the novelty of the sharply raked windshield, which projects the interior towards the rear. In the Cadillac the angle is barely hinted at, here it's loud and clear I couldn't say whether, due to this peculiarity, the Victoria Coupé was the first of its kind in the world, or whether my father gave it, for the first time, such a sharp rake. The doors are frameless, no heavy molding. The windows have practically no molding. The decoration denotes, as always, my father's tendency to divide the car in two, to make it more streamlined. The line of the front fender is extremely pure, it stretches out until it takes in the running-board and then rises light and soft towards the rear. The roof is extremely modern—very much in keeping with the definition of bodywork today, now that the period of square lines is over with. The overall style, the combination of details, make this car a prodigious example of elegance. There was an important client of ours from Switzerland, Mr. Ramseier, who was an importer in Paris and Geneva of Bentley, Rolls, and later Lancia. He used to come to Turin often, and my father set great store by his opinions. Well, as the story goes, just after the war when those pontoon-shaped bodies, joining the hood to the body side, were coming out, this gentleman, referring expressly to the Victoria and to the other cars of its ilk, exclaimed to my father: Pinin, you used to make greyhounds, and here you are making bulldogs! Mr. Ramseier was just horrified by the fundamental change in the structure of the jacket, I mean, the garment sewn around the mechanical parts. He belonged to the culture of greyhounds and didn't accept this bulldog culture."

The body of the Victoria clothed the 8-cylinder 2300 in the years when Alfa Romeo triumphed both on race tracks and in road races alike—one for all, the Mille Miglia [Thousand

In the rear the design is concluded symmetrically: the raised central part terminates by tapering into the license plate housing, while the fins take on relief and visual importance.

The streamlining is dominated in the front by the round protruding air intake which, in turn, gives rise to the rounded hood. Standing next to the car, Pinin Farina in a photo from the 1950s.

Miles]. The Super Sport and the Grand Sport were winning one race after another. And 1929 witnessed the start of the duel between Varzi and Nuvolari that was to whip millions of fans to a frenzy. In Modena Enzo Ferrari, at the end of that year, will found his racing stable. If the Victoria remained, in the Pininfarina version, one of its kind, it added a dream to the dream of whoever was lucky enough to view it at its presentation at the Milan Motor Show in 1933. "Nuvolari and Guidotti won the 1930 Mille Miglia racing a 1750 Gran Sport Testa Fissa," Luca Grandori writes in *Un sogno su quattro ruote* [A Dream on Four Wheels], "and Campari took fourth place in the Targa Florio with the same car. The Testa Fissa won the first three places at the Tourist Trophy, which took place at Belfast, in Ireland: Nuvolari, Campari and Varzi, first, second and third. The Testa Fissa version was the official 1750 with which Alfa Romeo took part in the races; twelve cars were built in all, in two series of six models each. The name comes from the peculiar construction of the cylinder head [*testata*]. In order to avoid blowing head gaskets—a not uncommon occurrence in extremely tough competitions such as the Mille Miglia or the Tourist Trophy—the cylinders and the head were constructed in a single block. The cars, moreover, proved to be much lighter and their performance rose to very high levels. In the 1931 Mille Miglia, in fact, Campari finished second overall, behind Caracciolo who won the race with a Mercedes with a 7-liter engine. Meanwhile Varzi and Nuvolari continued to fight it out. Varzi responded to the victory of his adversary in the epic Mille Miglia of 1930 by taking first place at Monza with a Maserati. Then Nuvolari entered the Targa Florio with a 1750

Leopold of Belgium at the wheel of the Ferrari 342 America conversing with Liliane de Rethy.

but, held up by a broken front leaf spring, had the victory snatched from him by Varzi, driving an Alfa Romeo P2. In 1932 Nivola took his revenge, winning the Targa Florio and, together with Campari, the Italian Grand Prix at Monza. The following year he did an encore in both competitions, finishing first in the Targa Florio and at Monza; but Borzacchini took the Mille Miglia. In the meantime privateers were using the 6-cylinder 1750 to good effect all over Europe, clocking up successes in all the important, and less important races."

So the greyhound that Mr. Ramseier, Pinin Farina's client-advisor, preferred over all the other bodies Pinin designed, was born in 1933, after years of resounding triumphs of Alfa Romeo and, not by chance, it bore the Alfa name. But then, in the late 1920s and early 1930s all the existing land speed records were broken—reality had far exceeded the early Futurists' wildest imagination. Luca Grandori continues: "The race towards the future was taking place on the sands of Daytona where in 1929 H. Seagrave drove the Golden Arrow at 231 mph. Two years later Malcolm Campbell raced the legendary Blue Bird at 244 mph and in 1932 he set yet another record: 252 mph… In the field of aeronautics 1930 witnessed a discovery that was destined to revolutionize the future of mass-transportation: an English engineer, Frank Whittle, patented the first jet engine. It was not until 1941, however, that the first turbine-powered airplane actually got off the ground. Also in 1930 Italo Balbo, with his group of twelve seaplanes, demonstrated the supremacy of the Italian aircraft industry, flying nonstop from Orbetello to Brazil. And in 1930 there were still other exploits. On the one hand the German dirigible Graf Zeppelin went around the world in 20 days; on the other, an American airplane broke the air duration record with a continuous flight of 150 hours, refueling the craft in flight. The American Admiral Byrd, setting out from the base of Little America, reached the South Pole with a trimotor after making important geographic discoveries in the Antarctic. And in 1931 the Swiss scientist Auguste Piccard became the first man to reach the stratosphere thanks to an air-balloon he had invented."

If the land speed records were reaching vertiginous heights thanks to the increased power of the propulsors, the example of aeronautics offered the possibility, if not to reduce the importance of raw power, certainly to put it to better use designing car bodywork according to aerodynamic principles. Pinin was sure of it, and in 1935 he designed and built the coupé Alfa Romeo Pescara, with a 2300 cc 6-cylinder engine. "The aerodynamic study is most evident in the Pescara," Sergio remarks. "That fairing of the individual wheels and of the narrow central body is clearly reminiscent of an airplane wing. A detail that was an absolute novelty for its time was the positioning of the headlights, incorporated in and protected by grills that repeated the design of the radiator grill. My father had a lot of faith in his own intuitions, but he also listened to the advice of Rosatelli, the

Ferrari 342 America Cabriolet (Leopold of Belgium), 1953

Pinin Farina and Leopold of Belgium the day of the delivery of the Ferrari 342 America Cabriolet.

engineer who had designed the fighter planes used by our army: the CR 32 of the war in Africa and the CR 42 of World War II. In fact 'C' stands for *Caccia* [Fighter] and 'R' for Rosatelli. And the engineer told him that it was wrong to fair everything in, it greatly increased the bulk of the main section. In an airplane a mistake like that would have been pure folly. So if on a wing, which is thin, an element juts out, then you have to fair in just the element, without modifying the rest of the wing's section. I myself was to carry this theory to the extreme, in my study of bodies for Ferrari: go down with the hood, fair in the carburetors that jut out but without covering them entirely which would have meant risking the dangers Rosatelli talked about. In the Pescara my father paid close attention to the aerodynamic function and worked out the most elegant form for it."

Pinin knew one thing perfectly well: of forms found in nature, the woman, with her intelligent physical conformation, was the most elegant model for the language of the plenum and the void, of bulges and hollows, vertiginous dives, soft and alluring curves. And in fact, where visual impact is concerned, the Victoria and the Pescara are both females of ancient beauty. But what exactly was going on in the world of women's and men's fashion in that period, when fashion plates of both sexes paraded in their prestigious automobiles? "Towards the end of the twenties long skirts were *de rigueur*," Luca Grandori notes, "as Paris—or, to be precise, Jean Patou—prescribed. Compared to the lengthening

Two air intakes, faired-in, follow one another on the hood, which points downwards, letting the fenders emerge.

Ferrari 342 America Cabriolet (Leopold of Belgium), 1953

In the Pininfarina factory in Corso Trapani workers apply finishing touches to the 375 Mille Miglia. The treatment of the rear window anticipates that of the Dino Speciale (1965), fitted between the two fins.

Roberto Rossellini and Ingrid Bergman in a photo from the 1950s. AFE, Archivio storico del cinema, Rome.

of skirts, which left bare no more than half of the calf, the cut of the hair was shortened to a lesser degree. Ladies did continue to prefer short hair, as in the 1920s, but the style changed, becoming wavy. Women with long hair had to adapt by trying out new solutions like the elegant little chignons worn at the nape of the neck. It would appear that the male model of the day was Edward, the free-and-easy Prince of Wales. People envied his natural class and his nonchalance… The elegant Italians of the day, when they weren't wearing their Fascist uniforms, emulated the style of the Prince, which was soon all the rage; for evening-dress they no longer wore tails but now preferred the tuxedo, a short black jacket with satin revers that the English call the 'dinner jacket.' During the day, however, the impeccably dressed man wore, without fail, a Borsalino or a bowler, patent leather shoes, a double-breasted coat with six buttons without half-belt and a dark-blue worsted jacket."

Alberto Moravia (1907–1990) gives us, in 1929, a merciless portrait of this Italian bourgeoisie in *Gli indifferenti* [The Time of Indifference], his first novel: "She took a tram that was going towards the centre of the city. Lisa must have been waiting for her for twenty minutes already, at the hat-shop where they had arranged to meet to examine the new Paris models. She sat in a corner by the window, turning her back as much as she could on the common people in the tram and looking out into the street. The pavements were crowded with a lively throng of workers of every kind going back to their homes; the cold February sunshine lit up their faces, reddened by the icy wind, beneath the worn brims of discoloured, shapeless hats, and their figures enclosed in overcoats that had turned greenish from long use. It was a white and hazy sunshine, and it spread itself generously over all these rags as though it wished to give them its blessing. One after another, the bright shops filed past, with their lettering in red, white or blue paint on the windows; electric signs hanging from the cornices, grey and dead, looked like ashen spectres; while the tram, many-coloured, crowded, vulgar as a merry-go-round, advanced slowly, shaking and clanking. From time to time, right under the eyes of the Signora, the shining, oblong bonnet of a car would come alongside with a swift movement, stop as if it were searching for a way through, with its big headlamps for eyes, then bound forward. She would see, behind a sheet of glass, a leather-clad chauffeur sitting in his place, his gloved hands on the wheel, and then, lolling on the leather cushions with a look of extreme satisfaction and

Even today the line of the Ferrari 375 Mille Miglia is extraordinarily elegant: note the retractable headlights, rather a novelty at the time, and the sharply raked nose.

a half-open eye lowered towards the crowd, some paunchy personage, or some delicate-faced, painted lady wrapped in voluminous furs. Then, unconsciously, she would sigh; for she herself would never be able to drive through the ill-dressed crowd in a powerful, impressive motor-car; her best years had vanished, her youth had fled in the shining car of her dreams. Then gradually these people that she envied, these ephemeral figures passing swift as arrows in their humming chariots, withdrew, also, from her fantasies and hopes, and, resigned, she continued on her way, with a kind of disgusted dignity, in the great painted wagon of iron and glass."

In 1926 Massimo Bontempelli (1878–1960), approaching Rome in a sports car, runs smack into the middle of a jumble of placid two-wheel carts. In the story "Da Roma..." [From Rome...] from his volume *Stato di Grazia* [State of Grace] he reminisces about it in language that mingles the real and the imaginary: "Approaching Rome in a super-fast Tipo Monza, we found ourselves meeting up with one cart, then two and three, then many—an immense multitude of carts going every which way. Nowhere in the world can you see as many carts as you see on the roads around Rome. They were in front of us, behind us, right at us; we were making our way through a dusty and shrill atmosphere of carts. They were covered by large spherical umbrellas of twenty colors; the wind was driving them like pagodas with full sails. Some emerge from side paths, or from the dry grass of meadows. Others come up out of ditches, right out of the ground. They come down like monkeys from telegraph poles, they pop out of yellow clouds and come down along the curves of the sky, plunging into the road in

The simple and compact passenger compartment of the Maserati A6 links up with the tail in a single section.

front of and all around us. At first the car dodged them, slipping through the cracks and scraps of the road like a ship in an archipelago. But then we gained confidence and changed our tactics: the Tipo Monza threw itself head over heels smack into the middle of this anthill of carts, come what may! Either it got by them, or it ran them right through. Even if some were sacrificed, there was still a horde of them left over, more than anyone needed. One or two were razed to the ground, like shades of carts long-dead. The others continued placidly making their way, the cart-driver asleep, the horse or donkey perfectly indifferent to the fate of its colleagues that had been swept away."

If it's true that in a few years time, and even more so in fact than in Bontempelli's fiction, the horse will be forced to give way to the machine, well, the sole protagonist of this dream of speed will be the power of the engines. As Sergio explains: "My father, in 1936, in his Aprilia Berlinetta Aerodinamica, hit 100 mph on the road from Rome to Ostia. Fully 22 mph more than the top speed of the standard, 1123 cc V4-engined Lancia sedan. He had succeeded in showing, just by modifying certain ratios, the importance of aerodynamic design. There's no doubt that, in 1936, the only way my father could test his intuitions was on the road, where he broke a record that the 'experts,' who counted on big engines, would not even have been interested in. But the news reports of the day exalted that race against time on the road from Rome to Ostia. The Aprilia Berlinetta Aerodinamica featured details of great stylistic finesse, from the placement of the air intakes to the design of the window frames. And it had been no easy job to create them. My father wanted, as far as possible, frameless windows. They were best from the aesthetic point of view, but were also the most exposed to pressure and depression when the car was in motion. One result was bothersome noise. So it was absolutely necessary to individuate the neutral point where the post was to be placed. But this meant taking the body apart and putting it back together again after the road-test that made it possible to pinpoint the defect. Much later on, thanks to the wind tunnel, the problem will be resolved, and with it the proper placement of the air intakes."

The Aprilia Berlinetta Aerodinamica is a milestone in the evolution of bodywork design. But beautiful cars are not only symbols of speed, they are also the means of scaling the peaks of society—or this, at any rate, is the opinion of the idle nephew of the two old maids, living off their embroidery, in the novel *Sorelle Materassi* [The Materassi Sisters] that Aldo Palazzeschi (1885–1974) published in 1934. "That day, after lunch, as soon as Giselda had gone out, Remo had to entertain his aunts with the account of some important subject. He was twenty-two years old now, it was time to think about a real job, he had his eye on certain possibilities that he didn't want to lose: specifically, the agency for the Florence area of a new car destined to be a surefire success. But to make contact with the head office and show them his worth, he himself

In Sergio's words, the Maserati A6 GCS from 1954 is one of the most beautiful cars Pininfarina ever made; in the center, the elliptical air intake with concave vertical grill and the Maserati logo, the trident.

had to possess a car worth thirty-five thousand lire. His whole future depended upon it." The maiden aunts adore their nephew and agree to bear with the expense. "Even more disproportionate to their way of thinking than it was to their actual possibilities… Listen… If it's true that you need the car to create a position, to open new avenues… Teresa said this but she didn't see what avenue her nephew could open by means of that car that opened all the avenues for him to hurtle down at breakneck speed… We've decided to buy it for you." Acceding, after that, to still other wishes of their nephew, the Materassi sisters sign their lives away. When the nephew, after ruining them completely, marries an heiress from across the ocean and takes off for America with her, the aunts are forced to go back to their embroidery to eke out a living. But they don't hold it against him; on the contrary, they often miss his contagious will to live.

If, however, young Remo in Palazzeschi's novel uses his career as an excuse to convince his aunts to make the fateful purchase, the truth is that agents of the major car companies, in those days, were people who really did count, and especially for their ability to interpret the wishes of their clients. Look at the Lancia agents, for example: Viara in Genoa, Guaragni in Rome, Antolini and Orsi in Bologna, Franco Bocca in Biella. As Sergio tells it: "A friend of my father's and of Lancia's, Bocca ordered the custom-built Astura that will bear his name. While the Victoria, the Pescara and the Berlinetta Aprilia were ahead of their time and heralded basic themes of future design, the Lancia Astura Bocca stayed put at 1937, as a super-luxury car that was

What strikes you is the sleekness of the hood, the sporting design of the roof, the highly innovative windshield.

Maserati A6 GCS, 1954

Working on prestigious foreign chassis, like that of the Bentley, meant taking a fresh look at the distinguishing features of these cars of renown, bringing them up to date.

The treatment of the entire front end of the Bentley Continental goes back to British canons; the rear is simple, with two small round tail lights that hint of fins.

closer to the pampered world of 'white telephones' typical of the rich bourgeoisie that people saw in the movies. It was an island of comfort at the twilight of an epoch that was about to enter the home stretch of the pernicious drift of the Fascist regime." And Lorenzo Ramaciotti, in *Solitaires*, his fine book dedicated to a selection of precious Pininfarinas, writes: "The car is imposing due to its extraordinary size, over 16 feet long, which is unmitigated by attempts to hide the dimensions but taken as a smooth monolithic mass... Unlike sports cars that call to mind motion and speed even when standing still, the Astura seems as if it will never leave the state of rest... The interior is practically an extension of the warm, safe atmosphere of the family easy-chair, transformed with ease into a means of transportation. The materials and the workmanship are refined but not ostentatious; in this particular car we have leather plaited in broad strips, with an eye for the tendencies of modern design, without excess and in good taste. To complete the refined richness of this object, the imposingly large convertible top is opened by a hydraulic servo-assisted mechanism—patented by Farina, as a plaque informs us—still in perfect working order."

There's no question that the construction of an Astura Bocca with its 2972 cc V8 engine required workers of consummate skill. "But soon, for elite bodywork of the sort made by Pinin Farina, the war was to be a severe test," Augusto Morello writes. "And all the more so since we know from Pinin himself just how intellectually and managerially modest his military clients were, not to mention his antipathy towards weapons generally, to the extent that he didn't even let his son play with the toy guns that had been given him as gifts. And so his production was limited mostly to accessories for the medical corps and transport trains... It was a miracle that he survived those five long years, especially during the German occupation when the small factory became a partisan munitions depot. A deferred consequence of those hidden weapons deposits came right after the war, with the enormous damage caused by the fire that totally destroyed the raw materials storage area, the most delicate part of the factory."

One particularly important event, however, was to herald the resumption of Pinin's international affairs, at a time when echoes of the war still hung heavy in the air. Pinin himself tells the story in *Nato con l'automobile*. "Sergio was twenty years old and going to the Polytechnic; he'd never taken a long trip, but I'd dropped him some hints on the subject. I already used to let him come to the factory, and I liked the idea that a student of mechanical engineering should also grapple with problems on the benches of a workshop. In October of 1946, along with some other cars we had finished an Alfa Sport 2500 Cabriolet and a Lancia Aprilia Cabriolet. I told Sergio that we had to take them to the Paris Salon... In 1946 Italy, as a vanquished nation, had been excluded from the exhibition... But I had a plan and I wouldn't have given it up for

At the 1954 Turin Motor Show Sergio Pininfarina and Renzo Carli examine the imposing hood of the Bentley Continental.

anything in the world… A car is made for the road, so the roads too, the public squares, can be a launching pad. Man doesn't live by conventions alone." Sergio drove the Aprilia and Pinin the Alfa. "At the entrance to the capital we stopped and cleaned the cars up… When we got to the Grand Palais where the Salon was held I set my plan in motion right away. Thanks to the complaisance of two *flics* [policemen] we managed to park the two cars right in front of the entrance to the show… Even the directors of the Salon, together with the car-makers, came to take a look—and all of them were old acquaintances of mine. '*Ce diable de Farina*,' they exclaimed, '*A ouvert son Salon personnel.*' They were cordial and even made cutting remarks about the severity of the *patrons* of the motor show… The next morning a paper published the news along with a photo of the *Antisalon du Carrossier Turinois Pinin Farina*. There were proposals, people made orders—it all ended up *en plein air*… The last day the secretary of the show, monsieur Mautin, told me: 'You've earned your place in the center of the Salon for next year.' For me it was as if the war, which didn't mean only bombings and strafings and ration cards, ended that day. It had lasted until then."

Italy just after the war, around 1946–47, witnessed the birth of a new form of theater, the musical comedy, which was definitively to assert itself in 1952–53. In his *Storia della canzone italiana* [History of Italian Song] Gianni Borgna writes: "Musical comedy made considerable use of songs, which helped make less provincial the theatrical texts and staging. It could boast such authors as Garinei and Giovannini, musicians Kramer and Trovajoli, stage-designer Coltellacci, choreographers Paul Steffen and Donald Saddler, actresses and actors Delia Scala, Isa Barzizza, Tina de Mola, Elena Giusti, Gianni Agus, Carlo Dapporto, Renato Rascel and others as well. Rascel in particular, in such shows as *Attanasio cavallo vanesio* (1952–53), *Alvaro piuttosto corsaro* (1953–54), and *Tobia la candida spia* (1954–55), launched a great many hit songs: from *La mia donna si chiama desiderio* [My Woman is Named Desire] to *Ninna Nanna del cavallino* [The Pony's Lullaby], from *Angolo di cielo* [Corner of Heaven] to *Dove vanno a finire i palloncini* [Where do the Balloons End Up]. It's strange that three of his best-known hits have to do with the sky or heaven." Renato Rascel purchased the PF 200 that Pininfarina displayed at the Turin Motor Show in 1953, custom-built on a 1991 cc V6 Lancia chassis. But our fine actor-singer could never have raced his car against the pony of his song, "through the pastures of the sky, in the chorus of stars of gold," since only the form of the auto rendered homage to the new jet aircraft, which had already become part of the collective imagination. The PF 200 was too slow for a jet, even if it looked ready to take flight any minute. Lorenzo Ramaciotti notes: "The streamlining, tapered progressively towards the tail, with perfectly smooth flanks, is emphasized in its length by the wheel base of the original car, but above all is dominated in the front by the round

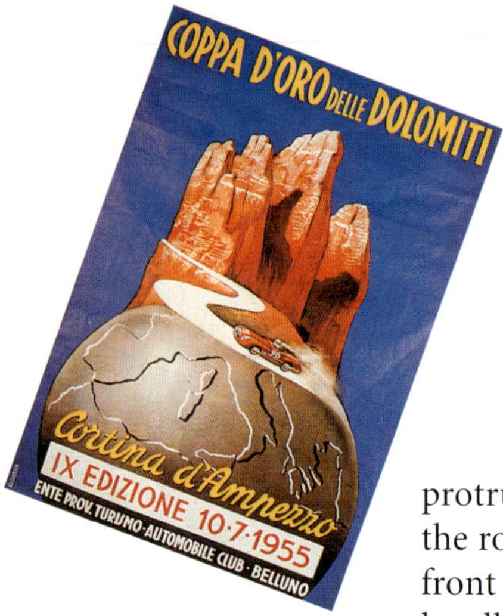

Poster of the "Coppa d'oro delle Dolomiti" race in Cortina d'Ampezzo, 9th edition, 1955.

Ferrari confesses to me, "Every car has a soul, and constructing them is like taking cocaine."

Guido Piovene

protruding air intake which, in turn, gives rise to the rounded hood. The other elements of the front are reduced to functional presences; the headlights and bumper guards frame and reinforce the principal presence. In the rear the design is concluded symmetrically: the raised central part terminates by tapering into the license plate housing, while the fins take on relief and visual importance, framing the large chrome guards of the bumper. The result is original and striking, even if the proportions are slightly forced—unquestionably an interesting piece of research." And Sergio comments: "Who else but an artist could ever have purchased the PF 200?"

It is one thing, however, to stand in front of a car, walk around it, touch it and desire to possess it. Rascel fell in love with the PF 200 when he saw it at the Pininfarina stand at the Turin Motor Show. It is entirely another matter to decide on a purchase when the line of a body is still in the mind of its designer and nowhere else. In this case, in order to seduce a possible buyer, it is necessary to create an image of something that does not yet exist. It is here that the rendering comes into play. As Sergio explains in his preface to *Pininfarina Ferrari 50 Figurini/Renderings/Esquisses*: "Apart from that of depicting the object, the rendering also has other, and perhaps more important, functions: to convince, to seduce, to arouse emotions, to stimulate the imagination… In this sense it is not only an expression of the visual arts, but also belongs to the techniques of persuasive communication; the favor and the possible inclination to make a purchase will also depend upon the suggestive force of the image. The function of the renderings is thus extremely complex. On the one hand they are the image and the interpretation of a designer's intention, the imaginary portrait of an object that will be; on the other, they are a means of persuasion and an instrument of sale. But it is precisely this second nature that lends renderings their autonomy and artistic dignity. The ability to

The Cabriolet 250 GT presents no particular innovation of form. The line is sober and flowing.

Ferrari 250 GT (Sadruddin Aga Khan), 1957

Detail of the driver's seat of the Cabriolet 250 GT: the body was made by Pininfarina for Sadruddin Aga Khan in 1957.

persuade in fact depends upon the expressive resources of the people who draw them, upon their mastery of techniques, and creativity, and ability to interpret the design through the filter of their own sensibilities and visual culture."

Especially in the case of one-offs, renderings proved to be essential when it came to winning the client over in advance. But what, in the end, will be more important, faithfulness to the intention of the designer or the persuasive sales pitch? "My father was leery about renderings. For someone like him who saw himself as a sculptor, a modeler of masses and of spaces, the rendering was a sort of mystification. His artistic morality was disturbed by the idea that the car generated at the end of the whole creative process would be different from that imaginary portrait that was the result of a mediation, of the translation into another language, which made use of conventions and codes of expression that were different from the ones in which he articulated his aesthetic discourse. Nevertheless, he knew that the rendering was the only way possible to speak about a complex product to people outside the creative process, the only instrument that could communicate the aesthetic characteristics, the novelty and the beauty of an object that, at that moment, was alive and real only in his imagination."

Leopold of Belgium in particular, among the people who were outside the process of creating car form, had a great passion for engines and machinery. He ordered from Enzo Ferrari a special cabriolet, on the chassis of the 342, with the engine of the 375: the 4523 cc V12. Ferrari assigned the design of the body to Pininfarina, which in 1953 produced the Ferrari 342 America Cabriolet one-off. The line of the side panels is taut. Two air intakes, faired-in, follow one another on the hood, which points downwards, letting the fenders emerge, to conclude the parabola extending beyond the line of the headlights. "Leopold used to take his winter vacations on the Côte d'Azur," Sergio recalls. "We showed him the renderings in the course of our visit to Brussels, and later on he came to see us twice, with Liliane de Rethy. The first time was while he was on his way to the French Riviera, and the second was on his return trip to Brussels. It took us six months to produce the car. And, after it had been delivered, Leopold informed us that the lateral suspension was very hard and that the conformation of the driver's seat, which he thought was poorly structured, gave him a backache. I was very young at the time, and I decided right on the spot to head for Brussels. So, when I got off the plane I took a taxi at the airport and gave the address of Laken Castle to the driver, who took me for a tourist wanting to see the royal residence first thing. When we got to the castle he drove me all around it to show me the best views, and then asked me: 'Where to now?' 'Inside,' I told him. 'That's impossible!' he exclaimed, astonished. 'This is the residence of the King, the flag is up so the King is there.' 'The King is expecting me,' I insisted, 'ask the guard.' Incredulous, he got out of the cab and went to talk to chief of the guard in charge of the gate.

A stratagem made it possible to get the Florida II homologated, when the law didn't allow doors open to the wind, but at the same time it was a great solution both aesthetically and for comfort as regards getting in the car.

The Florida II was Pinin's favorite, and he drove it himself for the last ten years of his life.

'The gentleman'—pointing at me—'says the King is expecting him.' 'Your name?' the officer asked me. 'Sergio Farina,' I answered. At the time we hadn't taken the name Pininfarina yet. (In 1961 Giovanni Gronchi, President of the Republic, will authorize by decree the change of the surname from Farina to Pininfarina).'Please, at the end of the driveway, the King will receive you at once.' The taxi driver sat back down at the wheel, in silence he started his engine and then, suddenly, he started talking to his cab, nearly as if it were a person, while he turned into the drive that led to the front steps of the castle. 'Remember what's happening to you, you old jalopy, you're about to present yourself before the King, could you have ever imagined such a thing?' Getting out of the cab I asked him what I owed him. 'I shouldn't take anything,' he answered, 'but I can't give up my fare.' I spent the day with the King listening carefully to what he had to say about the defects of the seat, we tested the car together. I was the King's house guest that night, and the next day, after a round of golf with my illustrious client, I took the wheel of his 342 and headed back to Turin. When the necessary modifications had been made, the Ferrari was sent back to the King who, this time, had no cause for complaint."

In 1954 a great cinema director, Roberto Rossellini, author of *Roma, città aperta* [Rome, Open City] of 1945, a masterpiece and the symbol of Neo-realism, ordered a car from the Ferrari works in Maranello. He wanted it as a gift for his wife, Ingrid Bergman. "Ever since he was a kid," Fernaldo di Giammatteo writes in *Roberto Rossellini*, "he loved cars (a passion he inherited from a daredevil aunt, the first female racing driver in the history of Italian competition). He dedicated days at a time to powerful cars. A friend of Enzo Ferrari's, he drove around in all the Ferraris he could get his hands on. With one of these in the summer of 1953 he went so far as to take part—with a reluctant Aldo Tonti as his co-driver—in the Mille Miglia. But he didn't get far; he was forced to drop out after a fleeting appearance. The car game enriches that world of games which fills his existence. Even his love for Ingrid, for his children (Robertino was born on 2 February 1950 and the twins, Isabella and Isotta, also called Ingrid, on 18 June 1952), and for his family (he never lost touch with Marcella, his first wife, and he followed the schooling of his son Renzo, whom everyone called Renzino to avoid confusing him with the musician) was all part of that joy of living which in him is second nature. He is always present, and absent, everywhere. He has everything, because he renounces nothing, but nothing ever totally occupies him."

Florida II (Pinin Farina), 1957

Roberto Rossellini and Ingrid Bergman were married by proxy in Ciudad Juarez in Mexico on 24 May 1950. The marriage kicked up quite a scandal. "In the United States tensions were unleashed that were practically unimaginable. A star like Ingrid Bergman, the wife of an honest professional and the mother of a daughter, for Hollywood represented artistic, professional and commercial certainty. Whoever dared to attack this certainty sooner or later would have to pay the consequences. Rossellini didn't understand this soon enough and committed, as the Americans saw it, some unforgivable sins. Worst of all, he had the gall to fall in love with Ingrid or, if you will, to requite the love that she showed for him. The reaction was immediate: puritanism and business formed an alliance." Ingrid turned her back on the America that had given her everything. "In 1950 she was 34 years old, with a shining career behind her, and two Oscars. Movie-goers all over the world remember her in Michael Curtiz's *Casablanca* (1942), Sam Wood's *For Whom the Bell Tolls* (1943), and Alfred Hitchcock's *Spellbound* (1945) and *Notorious* (1946). A Swede, she had been spoken of as the new Greta Garbo. She had been admired for her melancholy grace, capable of standing up to actors of the caliber of Humphrey Bogart in *Casablanca*, Gary Cooper in *For Whom the Bell Tolls*, Gregory Peck in *Spellbound* and, above all, Cary Grant in *Notorious*." Rossellini responds immediately to her fatal letter: "I am sure, absolutely sure that with your help I shall be able to bring to life a human being who, through hard and bitter experiences, will succeed in winning peace and in freeing herself of all egotism, thus attaining the only true happiness that is granted to humanity: the happiness that permits one to live in the simplest way possible, and the closest to creation."

But Fernaldo di Giammatteo warns us against judging their brief union too hastily, beyond its sentimental side that will conclude with a legal separation in 1957. "Some people maintain that the actress tied the director's hands, giving him little and taking a great deal away. This is true in part, especially if one considers the psychological schematism that Rossellini had to impose on his stories in order to make it possible for the actress to be herself. It is also true, however, that the clear face of a star, who forgets she is a star in order not to renounce being an actress, in a certain sense catalyzed—making them clearer and more visible—the profound tendencies (religiosity, critique of the bourgeoisie, moral nonconformism) that were stirring in the director's mind. If one has to speak of opposition—and there was always something oppositional in Rossellini's relations with his actors and actresses—one can conclude without hesitation that in this case the match was a draw. No one was defeated."

Guided by Roberto Rossellini, Ingrid Bergman will be the protagonist in 1949 of *Stromboli, terra di Dio* [Stromboli], in 1952 of *Europa '51*, in 1953 of *Viaggio in Italia* [The Lonely Woman], in 1954 of *Giovanna d'Arco al rogo* and *La Paura* [Fear]. And so, between one movie set and another, which occupied them both, and Rossellini's theater work (he directed Verdi's *Otello* and a production of *La figlia di Jorio*, a work by Ildebrando Pizzetti based on a tragedy by Gabriele D'Annunzio, for the San Carlo theater in Naples), the couple took a day off to visit the Pininfarina factory in Turin, to pick up the 375 Mille Miglia that Pinin had built to order for Enzo Ferrari. As Sergio tells it: "On the occasion of the Ferrari fiftieth anniversary celebrations, I saw that car take a victory lap on

Florida II (Pinin Farina), 1957

I recall that car-maker from Detroit, Olds, who used to live and sleep in his car. And I live in my Florida.

Pinin Farina

the circuit at Modena. Even today its line is extraordinarily elegant. There are very interesting features, like the retractable headlights, rather a novelty at the time, and the sharply raked nose. To get it at that angle, we lengthened the hood about thirty centimeters and moved the radiator air intake way to the front. Also the treatment of the rear window deserves notice. It anticipates that of the Dino, set between two fins. But where in the case of the Dino the design served the practical purpose of releasing the heat of the mid-mounted engine, with the 4523 cc, front-engined 12-cylinder 375, it's only an aesthetic touch. I received Rossellini and Ingrid Bergman personally and delivered the car to them, pleased with their compliments. I took them to lunch at the Cavallo di Bronzo. At that time there were two women at the center of everybody's attention. There was Sophia Loren, definitely an aggressive, pronounced sort of beauty, typically Italian but somewhat common. And then Ingrid Bergman, an ideal beauty, unlike any other, absolute. She had fallen in love with a man who was very intelligent but her diametrical opposite—the contrary of her innate and refined elegance. Anyway, she looked at him with a rapt expression while we talked over lunch."

In the early 1950s Enzo Ferrari, braced by the numerous victories on road and track of his flaming red cars, felt the need for a continuity of style in the design of his autos, which were now attracting the interest of private buyers all over the world. He knew that it was time to establish a preferential relationship with one single coachbuilder, who would follow the evolution of his machinery and complete its expression. Pinin tells us, in *Nato con l'automobile*: "Ferrari, through his trusted friend Carraroli, lets me know that he'd like to have a talk with me in Modena. I shoot right back: I'm more than ready to see him, but I'd like him to come see me, here in Turin. I wouldn't have minded at all if Ferrari had a look at my factory. They reply that Ferrari practically never leaves his place at Maranello. So he wouldn't come and he was very sorry about it, since he had an important proposal to make… We reached a Solomonic solution. We'd meet on neutral

Looking at the car you can see the serene equilibrium of volumes and the subtle tension of the dihedrals, the general perfection of the whole.

Florida II (Pinin Farina), 1957

Two photos of Pinin with his Florida II, on vacation in St. Moritz and near the factory in Turin.

ground, in Tortona, with few witnesses. The story went around that, since neither one of us wanted to fall off the horse, we simply *got* off the horse. But no one was further than we were from protocol, from political merry-go-rounds and bids for power. We both liked unadulterated realism, the language of the workshop and the background music of engines on the test beds. Accompanied by Sergio I met Ferrari at a *trattoria* in Tortona, and the conversation, which was very cordial, soon turned to that decision which each of us had already made for himself... The fact is that Ferrari and I both wanted to get the automobile away from the melodramatic, the monumental, even the visionary, and quit the decorative forever." Pinin realized right away that he'd have no trouble getting along with Ferrari, but ever since 1951 he preferred leaving to his son the technical and commercial relations with the House of Maranello and its undisputed Lord. But the foundations of that mythic stylistic continuity were also, and above all, due to the sensibility of Sergio, who was able to mediate the asperities of character of the two people he considered the "masters" of his life: his father and Enzo Ferrari. Thanks to Sergio's work the relationship with the Maranello manufacturer still demonstrates its vitality today, when, alas, the protagonists of that first agreement in the *trattoria* in Tortona, Enzo Ferrari and Pinin Farina, are no longer with us.

It was Sergio himself who became more and more convinced that, given his pact with Ferrari, it wasn't possible to extend the same design criterion to other car-makers. Pininfarina could no longer design for Maserati. And it was no simple matter, when relations with the Trident marque were excellent, and so were the results, as can clearly be seen in one of the last models, the 1985 cc, in-line 6-cylinder Maserati A6 GCS from 1954. "It's one of the most beautiful cars Pininfarina ever made," Sergio explains, "a real Maserati, there's no other name for it—it stands out from all the others. What strikes you is the sleekness of the hood, the sporting design of the roof, the highly innovative windshield." Lorenzo Ramaciotti describes it this way: "The front of the car presents three distinct spaces. In the center we have the elliptical air intake with concave vertical grill and the Maserati logo, the trident; and on the sides, separated by a groove that serves to set off sharply the low hood-line, we have the prominent fenders, topped by a slight creaseline. The simple, compact passenger compartment links up with the tail in a single section. The wrap-around windshield is distinctive, as it was rarely used on other makes of sports cars."

Ever since the 1950s the car has played an eminent role in the vast firmament of objects of use wether be luxury or economy. So what is the meaning of its mobile presence? "I think that today the automobile is nearly the exact equivalent of the great Gothic cathedrals. I mean, a great creation of its time, passionately conceived by unknown artists, consumed in its image, if not in its use, by an entire population that appropriates with it a perfectly magical object." This is how Roland Barthes (1915–1980) sees it in his book *Mythologies*. And then, in saluting the birth of the new Citroën: "The DS is first of all a new Nautilus." And what does he mean by that comparison? "All the crafts of Jules Verne are, really, perfect household nooks, and the enormity of their periplum adds to the happiness of their closure, to the perfection of their inner humanity. The Nautilus is, in this respect, the adorable cave: the satisfaction of

Sitting in the Florida II, clockwise from the driver's seat are: Andrea, managing director and general manager of Industrie Pininfarina S.p.A.; Lorenza, director of communications of Pininfarina S.p.A; Paolo, president and managing director of Pininfarina Extra S.r.l.; Sergio, president and managing director of the Pininfarina group.

reclusion attains its paroxysm when, from the bosom of this interiority without cracks, it is possible to see through a large plane of glass the infinite outside world of water, and so in a single act define the inside by means of its opposite. In this respect most legendary or imaginary crafts are, like the Nautilus, essentially a delightful reclusion. It's enough to give a man a ship to live in for him, without ado, to delight in the ship as a smooth round universe—a universe of which, according to nautical lore, he is god, and lord and master (the only master on board, etc…)."

And eleven years later, Jean Baudrillard in *Le systeme des objets*: "The domestic sphere alone, and taken as a whole (furniture, electric appliances, etc.), structured within the great opposition arrangement/environment, possesses, in its relative coherence, a positional value equal to that of the automobile. It's clear that, on the level of lived experience, the domestic sphere, with its chores, functions and manifold relationships, is much more important than the sphere of driving a car. But on the level of system, one has to admit that today the home constitutes only one of the binary poles of the overall system—the other pole being precisely the automobile. While recapitulating in itself the latent oppositions and significations of the home, the automobile adds to it a dimension of power, a transcendence—without, however, putting into question the system itself. In the automobile private everydayness takes on the dimensions of the world without ceasing to be everydayness. In this way the system is saturated with great efficacy without surpassing itself. Change of place is a necessity; speed is a pleasure. The possession of a car is still more: the driver's license, a sort of attestation of citizenship, is the credential of that strange movable nobility whose blue blood is compression or top speed. Is not the revocation of a driving-license experienced as an excommunication, a social castration?… Also the dynamic synthesis that the speed of the car offers is something radically different from any kind of habitual function. The motion, in itself alone, is cause for happiness, but the mechanical euphoria of speed is something else again—in the imagination it's founded on the miracle of change of place. Mobility without effort is a kind of unreal happiness, of suspension of existence…

In 1958 Prince Bertil of Sweden ordered the Ferrari 250 GT Coupé.

The custom-built car has a sober line: low-cut radiator grill, elegant design of the front air intake, quarter window close to the rear window.

The car is also a home, but exceptional; it is a closed sphere of intimacy, yet released from the habitual limits of intimacy, capable of an intense formal freedom, of a vertiginous functionality. Intimacy of the home is involution in domestic and habitual relationships. Intimacy of the car is an accelerated metabolism of time and of space."

Perhaps, as a privileged verification of this "accelerated metabolism of time and of space," famous coachbuilders try their hands at the most prestigious machinery. And so in 1954 Pinin produced the Bentley Continental. "My father always thought that a small coachbuilder should find a space of his own in the market and then act there as a leaven. And there's no question that working on prestigious foreign chassis, like that of the Bentley, meant taking a fresh look at the distinguishing features of these cars of renown, bringing them up to date." This view is confirmed by Lorenzo Ramaciotti: "The treatment of the entire front end goes back to British canons: vertical radiator grill, large centrally-positioned headlights, bulging detached fenders. The flanks are slightly enlivened by the volume of the rear fender and the roof section transition. The latter, which is small but has a large area of glass, with a three-piece wrap-around rear window, lends strength to the imposing volume of the monolithic lower body. The rear is simple and nearly obvious, with two small round tail lights that hint of fins."

In 1957 Guido Piovene (1907–1974) published *Viaggio in Italia* [A Voyage in Italy], an essay that hovers between news reportage and a survey of customs. Here he is in Modena, intrigued by "the special madness of the Emilia region. Here the mania for fast cars reaches its apex; but it's really a passion for technology, for precision, for the atavism of what had been one of our great centers of craftsmanship. Automobile factories, but very special ones: Ferrari, Maserati, Stanguellini—it is not just by chance that they are all in Modena. A visit to the factories presents one with an odd human panorama, made up of technical aspects, craftsman spirit, obsession with sports, and of personal extravagance. At Ferrari, for example, they make one hundred cars a year, and no more. The parts are made one at a time. 'Cars are persons,' Ferrari confesses to me. 'Every

Three paper seals of the 1930s dedicated to sports.

car has a soul, and constructing them is like taking cocaine.' These masterpieces of automotive technique are born in a climate of romantic excitement. The celebrated car-maker is moved by a recollection from his youth when, going through the Abruzzi region in a race, he was stopped by wolves. The conversation about cars joins up, helter-skelter, with remarks about women and semi-theological reflections on humanity and on sin. One would say that here the factory is the most suitable place for confessions and for letting off steam. After the reserve—sometimes excessive—of the industrialists of Liguria and Piedmont, here we go to the opposite extreme. Ferrari is annoyed: 'Why don't you ask me something indiscreet?' he says. 'About what? About God? The devil? The temptations of the flesh?' Not even the prices are fixed. How could they be, with cars endowed with soul? The only laws they can follow are the laws of sympathy. The price is a compromise between the value of the car and the face of the buyer."

Every car has a soul, Ferrari said to Piovene. And there was a soul for sure in the Ferrari 250 GT, bodied by Pininfarina in 1957 for Sadruddin Khan, son of the Sultan Mohamed Shah, Aga Khan III, the recognized spiritual leader of the Ishmaelites. Every year the members of the sect pledged their faithfulness to him in a highly unusual manner. The Sultan was invited to take his place on one of the pans of a rather sturdy pair of scales (they weighed over a hundred kilos), while on the other pan precious stones were heaped up, most of them diamonds, until their weight matched that of the Sultan's. This took place once a year, according to the custom. Sadruddin had a brother, Aly Khan, who was killed in a car accident; he was the father of Karim, the entrepreneur in the Costa Smeralda in Sardinia, the current Aga Khan. "There is no really great innovation of form in the Cabriolet 250 GT," Sergio explains. "Sure, there's no doubt it's a Ferrari, you can see that right away. The headlights are faired-in, incorporated into the fenders. The line is sober, flowing." Unlike his brother Aly, a playboy pursued by *paparazzi* of half the world who was to marry the "atomic" movie star Rita Hayworth, Sadruddin, after a thirty-year stint at the United Nations as High Commissioner for refugees, and Co-President of the Independent International Commission for humanitarian questions, founded in Geneva in 1977 the nonprofit institution called Belle Rive, and will personally head its program for the promotion and conservation of nature. He takes up the fight for the protection of life in all its forms. If Enzo Ferrari, the Drake of Maranello, was moved by the recollection of his youth when, during a road race in the Abruzzi, he was stopped by wolves—well, when he met this special admirer of his motor-cars, it must have occurred to him that the wolves were racing too. They were racing for the right to live, and it was quite right to reaffirm it in any way possible, even at the wheel of the 2953 cc, V12 250 GT Cabriolet.

"I recall that car-maker from Detroit, Olds, who used to live and sleep in his car. And I live in my Florida." Pinin tells the story in *Nato con l'automobile*. "There's a legend about a horseman who is permitted to depart for his last voyage, of no return, with his horse as his companion. If an adventure like that—many years from now, because I want to breathe a lot more dust of life—but if an adventure like that did happen to me, for my companion I think I'd choose the Florida." So cars didn't have soul just for the Drake, who heard its music in the song of his engines. Pinin saw soul in the completed

Ferrari 250 GT Coupé (Bertil of Sweden), 1958

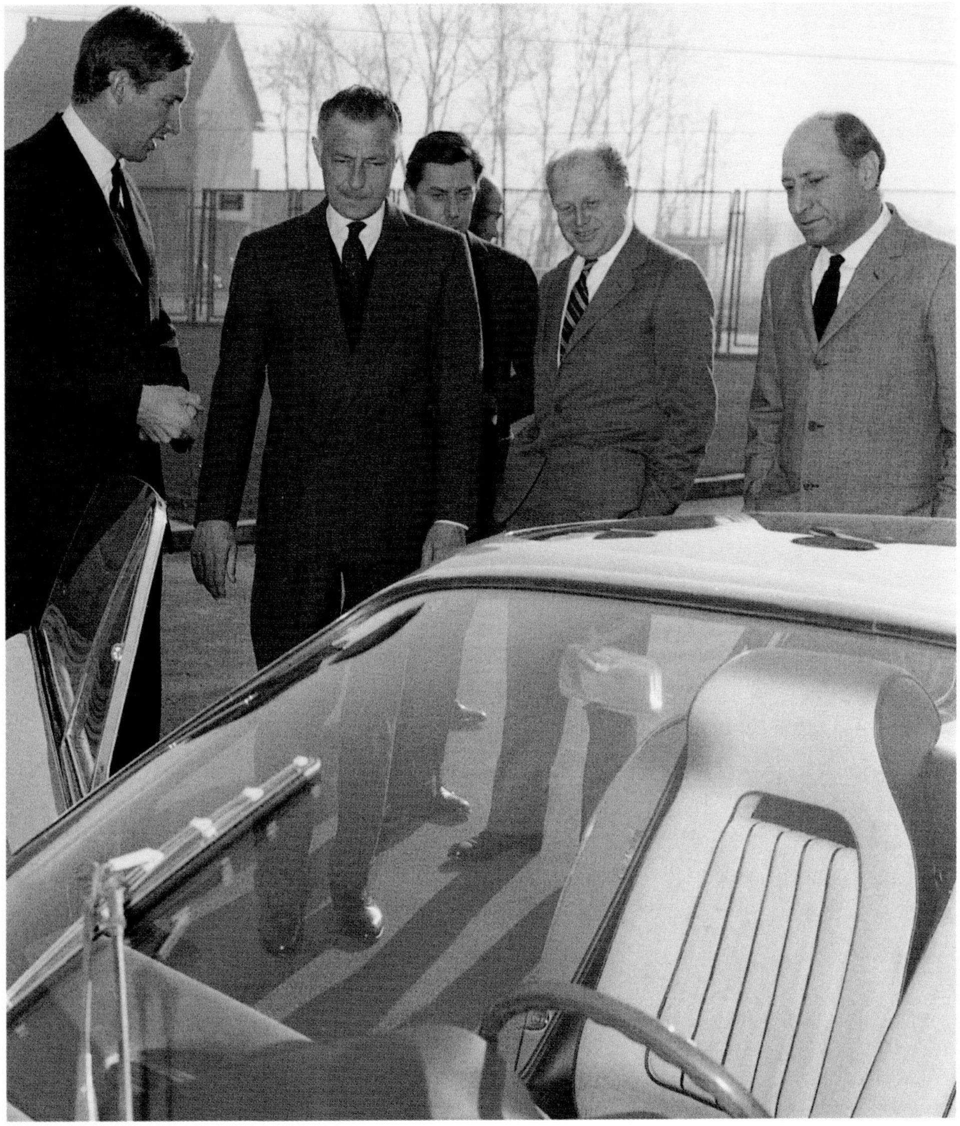

Sergio Pininfarina and Renzo Carli showing the Ferrari 365P to Gianni Agnelli, who ordered it from Pininfarina in 1966.

harmony of the form that corresponded best to the music of that song—whether the form was called Ferrari or by some other name. The Florida, for example, was a Lancia, a living being like the horse that accompanies the horseman on their last adventure. It was Pinin's favorite, and he drove it himself for the last ten years of his life. But why, exactly, that Florida II? "Looking at the car," Lorenzo Ramaciotti remarks, "you can see that the serene equilibrium of volumes and the subtle tension of the dihedrals, the general perfection of the whole, are not a result of chance but matured through a kind of purification." That sort of purification probably corresponds to the culmination of a process that went far beyond the corrections that distinguished it from the Florida I. Pinin was aware of this, and so he kept for himself the concrete realization of that magical point of formal equilibrium which had been attained. The Florida II, then, did not mean for him what the Flaminia Coupé, which was derived from it, will mean for some *nouveau riche* industrialist spawned by the economic boom of those years. For Pinin it was an archetype of any possible connotation that may be acquired by the Flaminia Coupé, its simplification. "It's a splendid coupé," Sergio emphasizes, "if you look at the car with the doors closed—just one handle on each side. But then the coupé turns into a four-door sedan, since the rear doors depend upon the opening of the front ones. This was a stratagem that made it possible to get the car homologated, when the law didn't allow doors open to the wind, but at the same time it was a great solution both aesthetically and for comfort as regards getting in the car. Then there are the panel creaselines, that give a play of colors with one single paint. And there's the saber-shaped decoration, the rear windshield-wiper, the extreme elegance of the fins. The Florida II is still part of our collection today. As a witness down through the years, it accompanies the most important events of the family."

In 1958 Prince Bertil of Sweden, Duke of Halland, son of Gustavus VI and of Princess Margaret of Great Britain, ordered from Enzo Ferrari a special coupé on a 250 GT chassis. Bertil was no longer a young man at the time, but in his youth he had been an athlete of some repute. In the early 1930s he came close to the world high-jump record; in Sweden he won the junior speed skating championship; he took part in motorcycle races with excellent results. In the late 1930s he was a diplomat in Paris and then in the Swedish navy, crowning his career with the rank of admiral. After the war he took charge of

The driving-position in the center makes the Ferrari 365P similar to a single-seater, to the racing prototype.

his country's exchanges with other nations as head of the delegations of commerce and of industry. The Royal University of Stockholm granted him an honorary degree in economics. Perhaps to recall his green years, Bertil decided to take the wheel of a car with the Prancing Horse logo. But he didn't want the car to be *too* sporty. "I remember Bertil, a real gentleman," Sergio remarks, "obsessed with simplicity. We produced a unique model according to that idea of his of 'not appearing,' of 'sober at all costs': the low-cut radiator grill, the elegant design of the front air intake, the quarter window close to the rear window to improve visibility. All-in-all the car was correct, simple, as our client wanted it. A 2953 cc V12 Ferrari which didn't reveal its racing-car nature at first sight but only to the person who actually drove it, discovering with pleasure its true identity. The yellow leather in the interior and the blue of the body are a reference to the Swedish flag."

Enzo Biagi published in 1971 *Il signor Fiat* [Mister Fiat], a biography of Giovanni Agnelli, *l'Avvocato* [the Lawyer]. Here is his description of Umberto Agnelli's attitude towards his older brother: "When he talks about Gianni he's affectionate and ironic, he enjoys himself: 'My brother has the defect that he takes an interest in too many things at once, so he's pretty unstable. He gets bored, because he's already thinking about something else, some other idea, he's afraid he'll lose something, he wants to do two things at once, if possible the one a thousand kilometers from the other.' The *Avvocato* admits: 'I'm always in a hurry, my schedule is overloaded with appointments that overlap geographically and qualitatively; at the extreme, it's a kind of schizophrenia.'" And this is what Umberto says about the big hurry his brother's always in: "He says to his driver: 'If I were afraid, I'd be dead by now.' He has a great time shaking off the police that have to protect him, taking curves on two wheels. An expert told me he could be a Formula One racing driver. But a Nuvolari—not a Fangio. No calculation, he just goes. To get where he's going, he wrecks the engine and risks everything. Gianni is very courageous because he's also something of an exhibitionist; but he really commits himself. He has a saying, whatever you do, do it well."

This is a curious comparison, that effectively refers to one of the greatest racing drivers of all time as a reckless car-wrecker. Aldo Santini, the author of *Nuvolari. Il mantovano volante* [Nuvolari. The Flying Mantuan], has a completely different opinion. Nuvolari did indeed take risks, but they were well calculated. "He didn't take curves on two wheels, he was no stunt man. Enzo Ferrari explains it well: 'I too, after the first few times I raced against him, began to wonder what was so special about the style of that lean and dour little man, whose value seemed generally to increase in direct proportion to the number of curves, which he used to call 'resources of the course'… Nuvolari attacked a curve rather sooner than my own driver's instinct would have dictated. But he attacked it in an unusual way, all of a sudden pointing the nose of the car at the inside of the

This side-view of the Ferrari 365P shows the lateral air intake and the scalloped body side.

The position of the gasoline filler, clearly displayed in the detail presented here, accentuates the reference to a racing prototype.

bend, right at the point where the curve began. Flooring the gas-pedal, naturally with the right gear engaged before making his frightful attack, he sent the car into a skid on four wheels, using the centrifugal force, reining it in with the traction of the driven wheels. For the entire arc of the curve the nose of the car shaved the inside curb, and when the curve ended and the car entered the straightaway it was already perfectly in line, with no need of corrections.'" Ferrari—as Santini writes—explains that Nuvolari will adapt his style to the technological evolution of the means at his disposal. "Later on, with the advent of independent suspension and when the tires were inflated at medium pressures, even Nuvolari could no longer skid in such an acrobatic way. But he still had something of his old style. He continued to attack the inside of the curve with the same decision as before, though now he'd correct his line by a few sharp turns of the wheel, no longer by systematically stepping on the gas. In any case Nuvolari's technique remained until the end a prodigy of instinct at the limit of human possibility and physical law. In fact in forty years of racing I saw only one other driver capable of it: Stirling Moss."

Santini and Ferrari, then, are in complete disagreement with Umberto Agnelli regarding the technique of Tazio Nuvolari. And as for Ferrari, he knew the "flying Mantuan" very well. A master-driver himself, no example of reckless driving—this was, unquestionably, also Gianni Agnelli's opinion when, in 1966, he ordered the 4390 cc, 12-cylinder Ferrari 365P. "We designed and produced the 365P, metallic gray with a blue and black band, for Giovanni Agnelli. A car that was extraordinary without a doubt," Sergio remarks, "suffice it to think that the driving position in the center made it similar to a single-seater, to the racing prototype. And to underline this similarity we placed the passenger seats towards the back, off-set with respect to the driver's seat. I drove this car myself, the feeling was unbelievable. The visibility was as commanding as it could possibly be, since the windshield pillars, from the central driving position, were confined to the extreme righthand and lefthand limits. Then there were the high fenders and the hood that dipped steeply and reduced the width of the front air intake almost to a trace, thanks to the mid-engined configuration. I would define the 365P a historic car."

Lorenzo Ramaciotti notes: "By 1966 the role of the coachbuilder had been clarified: on one hand the production of a small number of models, on the other creative consultation for the major automakers. Special cars, the custom-built one-offs for a single client, had become a thing of the past—not a remote past, but one that may be gone forever." This, then, is the deeper meaning of the definition "historic," referred to the 365P, perhaps one of the last one-offs to be custom-built by Pininfarina. From the mid-1960s onward the only one-offs will be research prototypes. Back in 1957 Guido Piovene, in *Viaggio in Italia*, had written: "The true custom-built car, unique and extravagant, today is dying. What is spreading in its stead is the mass-produced special, which is a sort of mean between the industrial disposition of our time and the vain disposition of men. The Turin coachbuilders work for the big industries, especially Fiat, but also for the Americans, furnishing that which the industries do not make—not single cars, but limited series of cars... The custom-made industry is part of the rigid, modest, thrifty and, at the same time, Pompadouresque panorama of the city. A city, as I said, that is severe and rococò, half accountants and half tailors."

Mass-Produced Cars

Detail of the front end of the Cisitalia 202, showing how the fender is higher than the hood.

The Cisitalia 202 and other models of the Pininfarina collection.

"I no longer recall which one of them, whether Bay, Barazzi or Reggiani, three engineer friends of mine at Pirelli, came over to my study that day—my office was no bigger than this room where we are now—to show me a material that had been used during the war for wrapping, protecting airplane fuel tanks. I'm talking about foam rubber. Anyway, this engineer says: 'See if you can find some new applications, in furnishings or in any other field you can think of. We'll give you working space in our Bicocca plant and the personnel you need.' And soon thereafter, sponsored by Pirelli, at the 9th edition of the Triennale I presented a furnishing that it was worth constituting a production company for. I wasn't able to buy into the company myself unfortunately, because I just didn't have a lira at the time." Marco Zanuso answers my question: How did your furniture archetypes come about and later give rise to a whole family of products? (In *Intervista* no. 14, July-August 1998.) "Until that time a seat had been more similar to a sculpture and the upholsterer who sculpted it had no idea at all of mass-producing the object. The 'Lady,' for the first time, created the possibility of assembling components that had been produced beforehand and, effectively, made the assembly line possible." Zanuso was 35 years old at the time. If the young designer wasn't in a position to finance his own company, there was no lack of entrepreneurs ready to embrace his design criterion. The Arflex corporation in fact was long identified with the Marco Zanuso name. As François Burkhardt writes in his monograph *Marco Zanuso*: "It was Zanuso who launched the typically Italian tradition of industrial designers trained in the schools of architecture, since Italy lacked a program that provided designers with specific university training in their field. Wresting away from the engineers part of a sector he wanted to make his own, Marco Zanuso is a prime representative of the self-taught designer who invents himself through continual contact with the practical, with work orders obtained thanks above all to his personal way of dealing with the problems of industrial production. Zanuso made it clear that the designer is not a decorator or a fitter of forms, but an inventor who is capable of thinking as a technician and a technologist, as a marketing expert and a consumer protector—indispensable as a guarantee of the manifold aspects of quality. Underlying Zanuso's activity are the foundations of humanistic thought, of that enlightened modernity which rests on the principle that construction is not a question of sentiment but of reason, of a positive mental condition dictated by the utility and by the

I'd gotten to a point in my work where technical know-how and pure form didn't count anymore; what counted was the ability to keep the two of them together…

Pinin Farina

The Cisitalia takes on the role of a symbolic watershed, with its whalebone radiator grill, its vigorous and elegant modeling, the hood lower than the fenders, the simplicity of the profile and the line sloping down towards the grill.

reality of the indisputable, fundamental elements of an attitude Zanuso still defends today."

In a photo from 1953 we see Marco Zanuso and Pinin Farina, together with Sergio, a young engineer and an attentive witness to the meeting. His father and Marco Zanuso were by no means decorators, fitters of forms. They knew about complexity, because design does not mean simplification—it means "project." But his father had also become an entrepreneur, even before typology, standardization and prefabrication became, as Burkhardt writes, "the themes of a will that now was clear—namely, the will to apply at last that which had been developed in the 1930s with ample references to the international debate in the USA, in England, France and Germany. Between

The Cisitalia 202 on exhibition at the Museum of Modern Art in New York. In the smaller pictures: the crate containing the car is loaded on the truck that will transport it to Genoa harbor for shipment to America; the Cisitalia 202 on the MoMA poster.

1946 and 1948 there was a feeling that rationalism, united with the drive towards the democratic renewal of the nation, could furnish the models and the programs for national reconstruction. What took place, then, was a reversal of trend. Where people had thought, on the basis of previous analyses, that it would be possible to open up a vast market, there now proved to be a qualitatively limited choice because the design that was proposed found no outlets. In the face of consumer rigidity, aesthetic dignity and systematic methodology are reduced to a supply destined for a very limited public. In the presence of cultural and productive innovations the following problem is posed more and more frequently: the objects can only be produced with semi-craftwork, and not industrial, methods. The dream of rationalism applied to design is realized only in single cases, which will become as many models—such, for example, was the policy of Adriano Olivetti and of his company in the 1940s." And Augusto Morello emphasizes: "The various peculiarities and scopes that induced some designers of that period to decide to become entrepreneurs have not yet been sufficiently studied. And yet there is no lack of examples: from Osvaldo Borsani of Tecno, to Gastone Rinaldi of Rima, to Gino Sarfatti of Arteluce."

Gino Sarfatti is the object of a critical reexamination written by Paola Proverbio in *Abitare* no. 380, January-February 1999. "Ever since the mid-1940s Sarfatti actively promoted an iconography of design that contained echoes of the most significant artistic languages of the period, from abstractionism to nuclear art. It was this innovative charge that made much of his production the equivalent of an archetype. In particular Sarfatti expressed an extraordinary sensibility and foresight in imagining new forms of lighting for living areas. In his work he incessantly experimented with the lighting devices available on the market… 'Three quarters of the lamps today are oil lamps. With the difference that the shape of true oil lamps was functional; it had to protect the quivering flame, and to have an exhaust pipe for the heat. But can that shape be adapted to electric light bulbs?' (R. Baldini, "I Personaggi di Gino Sarfatti," in *Casa Novità* [now called *Abitare*], 1 June 1961.) The starting point, however, was always the same, and in three parts: the bulb, the screen and the electric wire—in short the elements necessary for any sort of electric light… What Sarfatti did was to interpret the main technical components in such a way that each one took on distinguishing implications of expression. Following this principle, he subjected

An innovation with respect to the sports cars of the day is a greater concern with roominess, making for more comfortable seating.

the diverse typologies of lighting to a process of cleansing and formal renewal."

Pinin Farina and Gino Sarfatti, designer-entrepreneurs, were kindred spirits. As was Marco Zanuso—not an entrepreneur himself, he made the fortune of others. Even today, the objects of Marco Zanuso, Pinin Farina and Gino Sarfatti are considered archetypes, examples of the designer, on display at the Museum of Modern Art in New York. Pinin visited the MoMA in 1950; he recalls the occasion in *Nato con l'automobile*: "Half the rooms were occupied by a Frank Lloyd Wright exhibition, with relief-designs and huge panels... Those buildings appeared harmonic and in keeping with their environment. What struck me and convinced me most was the absence of decoration—no festoons, stuccos or affected detail, with a simplicity of motif that enchanted me... From the reproduction of these works you could intuit Wright's greatness. He created without hesitation, and without much confidence in the durability of modern cities, hemmed in by the old walls." Seeing Wright's works strengthened Pinin's convictions—away with all that reminiscence that still dominates car-body design! The body is the fruit of an advanced technological process and is in continuous development. It's form can be nothing other than the most significant interpretation at a given time of that process of coming into existence, foretelling its further and inevitable evolution. It expresses the same intentions of design that Gino Sarfatti and Marco Zanuso do, in different fields and with the various technologies at their disposal. Pinin Farina, Gino Sarfatti and Marco Zanuso truly understand the meaning of mass production.

The Cisitalia 202 participating in an Elegance Contest in 1947.

Cover of the magazine "Autoclub" of the Automobile Club d'Italia, 1959.

In the main room of the design section at the Museum of Modern Art dedicated to the automobile, the Cisitalia is on display as a symbol of that process of evolution whose necessity was interpreted so well by Pinin Farina. "I circled around the car and it was like circling all the way around my whole life… I saw the car through the eyes of someone who came later, with another age than the age of that car." Pinin had designed and produced the Cisitalia in 1947, and the MoMA was paying homage to his long voyage of coherence of design. Francesco Bernabò and Gino Rancati write about the Cisitalia in *Catalogue Raisonné Pininfarina* (1990): "The fact is that the formal approach to this car became the basis of the new aesthetics of the automobile, which in the following years was to influence the entire automotive industry all over the world." And Nino Balestra, in his essay *Intensamente Cisitalia*, writes: "The presentation took place at the Paris Salon of 1947; a month later, in November 1947, there was the *Mostra della Carrozzeria Italiana* [Italian Coachwork Show] in Milan. The distinctive term 'Cisitalia' was just something casually jotted down by Corrado Millanta. But it sounded good, and the word ended up in cursive script. The 202 has very sleek flanks. Pinin Farina succeeded in making this beautiful car also because he was given adequate running gear for the task, a chassis that will be exactly modeled to the line of the body, because around the doors and around the windshield all the tubes will be modeled to create a tubular cage. This was the first time in the world that a tubular framework was mass-produced (even if the production was limited to just over 160 cars, between coupés and cabriolets)." There's no doubt that critics all over

The Alfa Romeo 1900 Sprint Cabriolet is a full-fledged four-seater, very elegant in design, with well-modeled flanks and good-sized top and windows.

Alfa Romeo 1900 Sprint Cabriolet, 1950

The driver's seat of the 1900 Sprint Cabriolet.

The front end of the 1900 Sprint Cabriolet closely resembles that of the 1900 Sedan.

the world wondered, as they still do today, about the reasons for the symbolic meaning of the 202 sedan. The note of presentation in *Pininfarina, Sessant'anni* [Pininfarina, Sixty Years], published by the company itself, is quite interesting: "If the Cisitalia took on the role of a symbolic watershed, this depended on the coming together of various factors: the cosmopolitan flavor of its whalebone radiator grill, expressly related to the commonplaces of the official design of Detroit; its modeling that is vigorous and elegant but not too radical (the fenders are still carried over onto the flanks, albeit reduced to a role of ornamental low-relief); and, finally, its successes in the sports world that, together with the fact that over a hundred cars were produced, multiplied its visual impact." But Pinin himself comments: "To me the Cisitalia will always seem a riddle concealed in a mystery, which is enveloped in an enigma, something like the series of Chinese boxes, one inside the other. Or else I say that what gave me the idea was the hope for a better society, the confidence in our work… To obtain this I'd broken with a lot of very rusty old rules, which might have been OK in the days when to go for a drive people used to dress like bear hunters; I'd understood that the old forms had broken down… Anyway, I'd put

Alfa Romeo 1900 Sprint Cabriolet, 1950

The front end of the Nash Healey with its radiator grill marked by an ellipse that encloses the headlights and the radiator air intake.

The Nash Healey Spider with its top closed.

my ideas into practice without constraints. But I wasn't concerned with just some novel details, what interested me was the renewal of the architecture as a whole. The hood lower than the fenders represented exactly the opposite of the ones they were using especially in the American cars which, in that way, didn't give much visibility. The simplicity of the profile, the line sloping down towards the grill, the size of the doors reduced to a minimum, the windows no longer with those chrome frames and the absence of the rear pillar—all these elements and characteristics are topical, up-to-date. More than a few experts tell me that the sports cars they're making today owe a whole lot to the design of the Cisitalia. The overall vertical transverse dimensions, the design of the seats, the roominess, all respond to the demands of a touring car. And this means treating the passengers decently, where all too often they'd been packed into cars like so many crates. I'd gotten to a point in my work where technical know-how and pure form didn't count anymore; what counted was the ability to keep the two of them together, effortlessly or at least without showing the effort." So the Cisitalia represents a happy combination of the technical and the aesthetic. But then, some of the men on the staff of Piero Dusio, the industrialist from Turin who was "boss" of the 202, started to think that a more adequate engine would enhance the car's potential. Nino Balestra writes: "The idea was to mass-produce an engine with two valves per cylinder and around 85 horsepower, and one for racing with four valves and over 100 horsepower. Undoubtedly on that nice chassis an engine like that would have cut a magnificent figure and the car would probably have been unbeatable." Piero

The Spider in particular, so refined in its Turin edition, unquestionably represented the unattainable dream of a great many Americans.

The Nash Healey Spider parked in front of a billboard with Pinin Farina's face for the American advertising campaign.

Dusio, however, in the meantime had become infatuated with the idea of a Grand Prix car, and paid no more attention to problems regarding the mass production of the Berlinetta 202.

In their essay "Storia del disegno industriale" [History of Industrial Design] Giampiero Bosoni and Andrea Nulli write: "Alberto Rosselli, under the editorship of Gio Ponti, in 1951 began writing a column in *Domus* titled "Disegno per l'industria" [Design for Industry]; later on, in 1953, also in *Domus*, Rosselli published his "Manifesto del disegno industriale" [Manifesto of Industrial Design], and then, in 1954, he founded the review *Stile Industria* [Industry Style], of which he was editor; in 1954 the Compasso d'Oro prize was instituted, and in 1956 the ADI (Association for Industrial Design) began its activity. The stock of well-designed objects was still very limited; real meetings between industry and the culture of design were still rare and, above all, absolutely marginal." This culture, according to Arturo Carlo Quintavalle and Renato Fusco, as François Burkhardt notes, was influenced in the post-war period by the development of the plastic arts, "displaying the sources employed, the trends followed and the results obtained by the personalities of design: Sottsass (pop art), Castiglioni (ready-made), Mari (op-art), Munari (constructivism), and Sambonet (abstractionism), to name but a few. They demonstrated that it was not a question here of a simple transfer of types, but rather of a genuine personal elaboration, which sprang from very diverse roots, in a specific and individual way." And Burkhardt goes on to say: "By the term 'industrial design' is meant the mass production of objects, which are thus all identical and repeated in a number of copies; this differentiates

Audrey Hepburn and William Holden on the set of the film "Sabrina," directed by Billy Wilder. AFE, Archivio storico del cinema, Rome.

In this vintage photo, Pinin Farina and Mr. Healey, president of Nash Healey, in the Spider designed by the Italian coachbuilder.

itself from the concept of uniqueness and from the meaning of design that holds sway in the arts or the crafts. The activity of the industrial designer is therefore linked to an idea of mass production that tallies with a standard model corresponding to the production processes typical of industrialization. The aesthetic value of the mass-produced industrial object depends, then, on its configuration, which will make it possible to recognize its function, its construction, but also its origin in a specific industrial process. The function of an object of this type has to be linked up with its use, taking into consideration all the technical, ergonomic, psychological and representational aspects associated with its perception."

In post-war Germany the culture of design is linked up with the tradition of the Bauhaus, the mythical school founded by Gropius. As K.S. Leuschel writes in *Design. Sviluppi in Germania* [Design. Developments in Germany]: "In Ulm in 1949 Inge Scholl conceived of a *Hochschule für Gestaltung*, a high school for vocation training, which was actually built in 1953, with funds provided by an American foundation and subsidies from the German economy. It was founded upon the pedagogical principle of specialized training on a universal base, in union with a general culture in keeping with the times." The foundation that made it possible to open the school bears the name of Hans and Sophie Scholl, Inge's brother and sister, who had been killed in 1943 for taking part in a group opposing the Nazi regime. "Max Bill became the first headmaster of this high school of art, crafts and design, which until it closed in 1968 was to be of vital importance for training in the entire sector, even winning

The body of the Lancia Aurelia B24 is tightly reined in, yet the tautness does not detract from but, on the contrary, exalts the plasticity of the design.

Lancia Aurelia B24, 1954

In 1954 with the B24 Pininfarina invents the category of the fast touring car.

In the Italian cultural context then, given the lack of schools in the institutional sense, we witness the advance of the self-taught designer, who deals on his own with the strenuous task of approaching and working with the industrial sector: Marco Zanuso is unquestionably one of the prime examples. The designer-entrepreneur, on the other hand, is obliged to mass-produce a limited quantity of objects which, due to the modesty of demand, large industries have no interest in producing. We must add that the quality of the form cannot by itself produce greater sales in a broadened consumer market. And so Gino Sarfatti begins to mass-produce his forms of lighting devices, while Pinin Farina designs and mass-produces a small number of specials for the big car-makers that do not find it profitable to invest in niche products. In order to do this however, both designer-entrepreneurs, in their different fields, have to equip themselves to cope with the new production demands. "At the end of the 1940s," Sergio explains, "Alfa Romeo was torn between

itself an international reputation. It may in fact be considered the first international project of the newly founded Federal Republic of Germany. From the very beginning the school welcomed students from many other countries, including Japan and the United States, and attracted the interest and also the co-operation of young teachers who defended and spread its doctrines with enthusiasm and zeal."

Rendering and plan of the Lancia Aurelia B24.

Lancia Aurelia B24, 1954

And if to the eye the B24 seems already in motion when in fact it is standing still, this is due to the fact that its look weds perfectly with the designer's intention.

internal structural questions and a production policy based on large sedans, cabriolets, coupés with 2500 cc, 6- or 8-cylinder engines, which were just marking time on the market; then, management came up with the idea of a medium- to low-powered sedan and decided to bet on the 4-cylinder 1900." Luca Grandori writes in *Un sogno su quattro ruote*: "Until 1950 the Alfa Romeos were produced in two separate workmanship procedures carried out entirely by hand: the chassis on one hand and the body on the other. The 1900 marks the debut of the first Alfa Romeo monocoque chassis; at the same time, the first assembly line in the Alfa plant at Portello made its appearance. The United States in this regard were far ahead: the Americans were using completely automated production cycles, with machines capable of carrying out nearly all the operations. In Italian factories it was a different story. Here the pressing departments were still reminiscent of the mastodonic industrial architectures of Fritz Lang's celebrated film *Metropolis*, and yet from that mass of gears a unitized body emerged, streamlined and light… The production figures at Portello began to carry a certain weight: 1220 cars in 1951, 3107 the following year and 3115 the year after that." But, Sergio notes: "The 1900 sedan was a great success also because, given the same cylinder capacity as other cars on the market, it was the one that ran best of all, the fastest, thanks to an engine of 1884 cc, double overhead camshaft, 4 cylinders, 90 horsepower at 5200 rpm, as powerful as the previous 2500 cc engine of the large sedans. It was launched with no small fanfare, proposed as a car for the family and for winning races too. Even mom can drive it, the advertisement ran." There hadn't been much attention paid to women drivers at that time. Thus the ad for the 1900 aimed to give them their due, giving the lady of the house (there hadn't been talk yet about "working girls") a chance to imitate at least the driving style of Leila in the story "Fiori finti" [Artificial Flowers] by Corrado Alvaro (1895-1956): "When I met her she was driving her car in that negligent and virile way of hers, not with the studied diligence typical of women drivers." And Sergio: "Encouraged by the great success of the sedan, Alfa produced the chassis to be entrusted to the coachbuilders. So we produced a stock of about a hundred Sprint Cabriolets, a full-fledged four-seater with a very elegant line—well-modeled flanks and good-sized top and windows. The front end is quite similar to that of its sister model."

Between 1949 and 1950 there was renewed interest in the work of coachbuilders on the part of constructors and the public alike. "As if we'd become famous!" Pinin remarks. "And my colleagues had the same impression: Bertone, Ghia, Fissore, Canta, Allemano, Moretti, Francis Lombardi, in Turin and in Piedmont, and elsewhere, Boneschi, Borsani, Zagato, Colli, Touring." And this interest was not limited just to Italy or Europe. "Early in 1949 Mr. [George] Romney [assistant to the president of Nash-Kelvinator, which was later to become American

Film shot for a 1955 documentary: the assembly department of the factory in Corso Trapani.

The B24 on display at the 1955 Geneva Motor Show.

A line is OK when it is taut and correctly bound up at the extremities. A car can be rounded without being soft.

Pinin Farina

Motors] came to Turin, as he'd said he would. I had the impression of an extremely intelligent person, surprisingly modern in his approach to technology and to society… In three days of negotiations we came to a general agreement that included the designing of new Nash cars for large-scale production. We also aired the possibility of building, in our plant, Nash sports car bodies with engines to be developed by the English. I was flattered by these proposals, but I was also well aware of the inherent difficulties. It was imperative that the aesthetic ends should not be swallowed up by the industrial mass production on such a large scale." Pinin will be the protagonist in the United States of the advertising campaign for the Nash production that bears his name. "The idea was this: all over the country they were going to put up 4500 billboards 6 meters by 4 with an enormous portrait of me and no writing at all. The press agents were absolutely sure that this square Piedmont face of mine, looking out at the people all over America, would have aroused their curiosity and led to all sorts of guesses as to who in the world I was. And once these guys get started there's no stopping them. After a while the posters on the billboards were to be replaced by new ones with the words: 'This is Pininfarina, the coachbuilder, who has designed the most beautiful cars in the world.' Sure, it was pure conjecture, but repeated several thousand times it became an act of faith. It was an example of declamatory graphics, an organized system of persuasion. They thought it was valid to carry out this sort of 'rape' of the crowd, which in some ways reminded me of the rape of the Sabine women. On the eve of the presentation of the models the cycle was

If Bruno Cortona (Vittorio Gassman) in "Il Sorpasso," is a significant figure in the story line of Risi's film, the streamlined body of the B24 anticipates themes that go far beyond that story. AFE, Archivio storico del cinema, Rome.

VEGLIA
GIRI x 100
BENZINA
LUCI
RISERVA
RISCALD

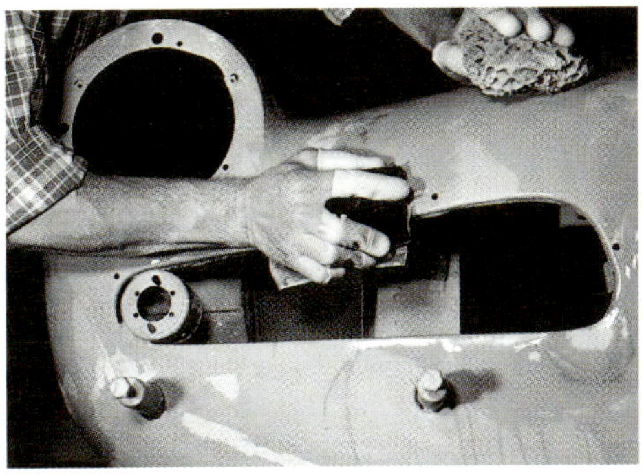

Two steps on the production line of the Giulietta Spider: with this car the industrial production of Pininfarina begins.

supposed to close, like a snap-shut hinge, with a third wave of my portraits bearing a supplement to the previous announcement: 'And now he has designed the new Nash cars.' The estimate for the campaign, including some filmed documentaries and radio and TV interviews, came to 5 million dollars. Had it ever happened before that the advertising for an automobile concentrated on the person who had designed the body?" And it was surprising indeed that a big American car-maker, for the first time in automotive history, to promote and sell its product played on the form and the paternity of its style. No longer, then, the strategic calculation of persuasion based upon big cylinder capacity and the relative number of cylinders—in short, of the horses under the hood. "On my arrival in New York harbor journalists and TV people boarded the *Ile de France* like pirates, shooting me from the front and from the back using the skyscrapers as stage sets. When I got off the ship I saw the first Nash billboards with my super-king-sized image, and a lot of writings such as 'Poetry in motion = Car designed by Pinin Farina' and 'He fills space with steel, wood and paint: the result is a painting.' The president of Nash, Mr. Mason, and C.H. Doss, one of the vice-presidents, were there to welcome me. At the Waldorf Tower there were other perils—photographers, cameramen. There was a press conference where they barraged me with questions. The next morning a reporter from the *New Yorker* showed up, and in the interview some reference was made of electricity in the air. I really do think those walls were magnetic, radiant surfaces."

The standard Nash cars, produced in America to Pinin's design, were a huge success.

Giulietta: the name is fresh, lively, quick off the tongue: the "fair maiden," as Pinin himself will call it, to distinguish it from the "fair ladies," famous and famed queens of the road.

The Alfa Romeo Giulietta Spider designed by Pininfarina in 1955.

Alfa Romeo Giulietta Spider, 1955

James Stewart visits the Pininfarina factory in the 1960s.

Pinin Farina at the wheel of the Spider with the pilot Guidotti at his side, at Balocco in the summer of 1962.

Pininfarina, for its part, built the Spider and the Coupé on Nash-Healey chassis, 500 cars altogether. A characteristic of both models was the radiator grill marked by an ellipse that enclosed the headlights and the radiator air intake. If the Nash produced in America moved the company up from eighth place to fourth in the classification of the most important US corporations, the Spider in particular—so refined in its Turin edition—unquestionably represented the unattainable dream of a great many Americans. And what else but a Spider could Billy Wilder have chosen for his film *Sabrina*! It is driven by William Holden in the role of David Larrabee, son and heir of a millionaire family. David prefers it to the dark sedans of the family stable. The young man falls in love with Sabrina, the chauffeur's daughter. He notices the girl for the first time when she returns from Paris, where she has been studying for the past two years. This *coup de foudre*, in the opinion of his brother Linus (played by Humphrey Bogart), threatens to compromise a marriage engagement that was of far greater advantage to the family. Linus will try everything to dissuade his brother from this passion, which he does not share, and make him see reason. In the process, however, he exposes himself to Cupid's arrows. While his brother drops out of sight, it will be Linus who falls irremediably in love with the girl and finds a way to make her fall in love with him too. It appears that Humphrey Bogart was dissatisfied with a part that was made to order for Cary Grant and, ill at ease, caused continual tensions on the set. In the film, Billy Wilder treats his portrait of American high society with clear-eyed intelligence, especially where it comes to class discrimination, expressed to perfection by Sabrina's father: "Life

"This car has a fundamental significance for us," Sergio explains, "it means the effective transition from craftsmanship to industry."

is like a car: some sit in the back, and some in the front. And in the middle there's a pane of glass." It's like that Spider, which takes its place alongside the dark sedans; it has no discriminating pane of glass, but it's still just a dream for the populace. The film will be a great popular success. But if in 1951 in the United States someone who purchases a Nash-Healey Spider or Coupé with its 4138 cc, in-line 6-cylinder engine, chooses the *crème de la crème,* and pays the price, in Italy someone who chooses the Lancia Aurelia B24 doesn't just make a statement about his personal fortune but confirms a precise way of understanding the automobile. With the B24 Pinin invented, in 1954, the "fast touring" category, at a time when his B20, the 1991 cc Aurelia coupé, was already showing such sporting merits that it was actually taking part in racing competition. Pinin tells us: "Among the over 300 cars that the evening of 28 April 1951 took off from Brescia towards Veneto, in the Mille Miglia, there were four of our B20s (aboard the 12-cylinder Ferraris there were Marzotto, Villoresi, Ascari, and Serafini). The B20s were driven by the Bracco and Maglioli, Valenzano and Maggio, 'Ippocampos', and Grolla and Monteferrato. At the Ravenna check-point Bracco was fourth, and in Pescara he was already second behind Villoresi. Then on the stretch from Rome to Siena the two drivers from Biella beat Bornigia's 3.5-liter Alfa by over two minutes and Villoresi's Ferrari by more than five… Bracco came in second overall at the finish line in Brescia, he was thirsty as the devil since he hadn't had a drop to drink for the entire 807 miles. Bracco was a rain man, like Nuvolari, and if it had rained during the race he might have had an even more electrifying surprise in store. In that

Alfa Romeo Giulietta Spider, 1955

The Giulietta Spider fills a space that has always been empty, this side of the mythic Ferrari and Maserati grand touring cars: for the price of a tame Fiat 1400 you can drive a thoroughbred that does 100 mph easily.

same year his B20 won the 6 Hours of Pescara and triumphed in its class at Le Mans."

The B24, then, is a "mass-produced special." As Marta Boneschi observes in *Poveri ma belli* [Poor but Beautiful]: "The experience that comes closest to championship races is the trip you take in a custom-built car, the queen of desires, which represents the best in terms of mechanicals, top speed and comfort. If it's out of reach, who cares? The Italian car industry won't be stopped by details. It tries, tries again, and invents, until it comes up with a solution within almost everybody's reach… Car fans get together once a year at the Mille Miglia, with its roaring autos and idolized champions, even though more and more frequently in the 1950s the race ended in tragedy. But just this is what makes the drivers heroes and supermen, at least in the eyes of their public."

Tazio Nuvolari died in Mantua in the summer of 1953. He had been suffering for some time from exhaust gas poisoning, which this fearless knight had contracted on the field of so many battles, pressing his adversary, forcing him to relent and give way. "In 1954 Alberto Ascari participated for the fourth time in the Mille Miglia with a 3.5-liter Lancia. At Desenzano a reckless spectator blinded him with one of those lights that was used to see the numbers painted on the cars to identify the drivers. It was a disaster, Ascari lost control, skidded, and in the accident that ensued one spectator was killed and another injured. In the same race, Nino Farina broke an arm and Eugenio Castellotti pulled out with mechanical problems. Ascari won, surviving both troubles and rivals." Nino Farina was Pinin's nephew, son of his brother Giovanni, who had taught Pinin the coachbuilding trade. "In May of 1955 Ascari was the victim of a frightful crash into the sea, but four days later he was at Monza ready to race. 'The dive,' he declared, 'is all over with. For a racing driver only tomorrow counts. In Montecarlo my lucky star protected me.' That star had done its duty for the last time. Alberto was killed during testing on the Monza circuit, exactly thirty years after the death of his father, Antonio, who had lost his life at Monthlery in France." How it happened is not perfectly clear. Did Ascari go off the track to avoid hitting a maintenance man who suddenly appeared in front of him? Did he blow a tire? The epic of the Mille Miglia was drawing to its tragic close. "In 1955 Eugenio Castellotti raced with the Ferrari backup car and had a great deal of trouble finishing the race… In the Mille Miglia the previous year Castellotti went down in the history of motor-racing, managing to win the

Advertising campaign for the Alfa Romeo Giulietta Spider.

Pinin Farina and his son Sergio with Sir Leonard Lord and Sir George Harriman, respectively the president and vice-president of the BMC, in 1958.

Sir Leonard Lord had understood before anyone else that a family car, to be successful, had to offer plenty of usable space.

race even though he had driven without his protective goggles, and under driving rain to boot. He had forgotten them in his haste at a refueling stop. He took off like a shot, and when he realized what he'd done he knew perfectly well that going back for the goggles meant losing the race. He crossed the finish line with swollen eyes." Castellotti was killed in Modena in the spring of 1957, during testing, on the track near Via Emilia. Perhaps a problem with the transmission, or with the clutch. From the pits the mechanics heard distinctly, in the sound of the engine, the driver's repeated, desperate attempts to downshift, and then the tremendous crash into the enclosure wall. "The following autumn he was to have married Delia Scala, the beautiful and bubbly soubrette. For the rest of the decade the Italians wept over the story of their broken love." And in the same year, 1957: "At Guidizzolo, on the road between Brescia and Mantua, the car driven by Alfonso Cabeza de Vaca, Marquis de Portago, is racing at 170 miles per hour; the Marquis is the fiancé of Linda Christian, veteran of many marriages but still a beauty. The front tire blows, the car skids off the road into a crowd of spectators. In an instant eleven people are killed, including the driver and his co-driver, the American journalist Edmund Nelson. The winner is Piero Taruffi, who's been trying to win for thirty years." Alfonso de Portago's accident puts into question all the road races, and among them especially the Mille Miglia, which will be the first one to be suspended. But the racing fans are disappointed. They miss the vigil, the being up until dawn waiting for their heroes to go by. The appointment once a year was the feast, the ritual to be observed without fail. The lovers of road racing all over Italy know very well that there

With the Austin A40, the body is no longer conceived as a smooth surface but rather as an architecture consisting of lines, planes and dihedrons.

will be no more emotions like that, once they're crowded together in the stands of the protected circuits.

And that kid in Fellini's *Amarcord* will no longer be able to dream of winning the heart of his beloved Aldina, whizzing past her window at the wheel of his hot rod. In reality, the busy streets are busier all the time, but with young toughs and notorious braggarts showing off their wheels—especially one in Rome by the name of Torello. "Everything was going badly that summer." So begins the story "Scherzi di Ferragosto" [August Holiday Pranks] from the *Racconti romani* [Roman Tales] by Alberto Moravia (1907-1990). "And, the day of the August Holiday, I found myself in Rome without friends, without women, without relatives, alone. The shop where I was a salesman was closed for the holidays, otherwise, out of desperation, just for the sake of company, I'd even have resigned myself to the summer sales, underwear, stockings, shirts, all old stuff. So, that morning of the fifteenth, when Torello came beeping his horn under my window and then invited me to go with him to Fregene, I thought: 'He's unpleasant, he's downright obnoxious… but better him than no one.'" More than his arrogance, which he owes to the money in his pocket, the real problem with Torello is his lack of even the most elementary sense of propriety. "Torello had a car with a custom-built body, green and low, of which he was proud as could be. Still inside the city, just after Saint Peter's, he started racing it like crazy: ninety, a hundred, a hundred and ten, a hundred and twenty [kilometers] an hour. I told him: 'Slow down, where's the fire?' And he answered me by giving it more gas. So in a flash we shot past the

Working together, in 1958 Sergio Pininfarina and Alech Issigonis produce the A40, an auto that already anticipates future hatchback cars.

The Ferrari 250 SWB, as a technical-aesthetic synthesis that was attained at a precise moment of a process of evolution, of Ferrari in particular, and of Pininfarina, constitutes itself as an archetype.

Madonna di Riposo and headed down the Aurelia. Because of the August Holiday the road was full of cars and for Torello it was a point of honor to pass them all, without honking his horn, without looking to see whether the way was clear; with his head down, just like a bull [*toro*]. Finally we hit a straightaway and down at the end we saw a big American automobile, going pretty fast itself, black and sparkling in the sun. 'Now we'll pass that one too,' said Torello, and hit the gas. It was a more powerful car than ours, but the man at the wheel was driving carefully, with regularity; there was a woman at his side. Torello came up behind, there was a bend in the road, he drew level and saw the woman: blond, with a round face, eyes of black velvet, a sly and slightly depraved expression: a big cat. The man appeared to be short, with a nose shaped like a bell clapper. He was driving with a cigar in his mouth, his sports shirt half open, his hairy arms on the wheel. Torello yelled, 'Bye-bye beautiful blond!' and she turned and smiled at him. In that very moment a truck big as a house came around the bend, and the man with the cigar, ready for it, swerved his big American car off to the side, followed just in time by Torello. The man with the cigar gesticulated and took off again like a flash." The total lack of propriety will, sooner or later, get the slow-witted Torello into serious trouble. And this also speaks volumes about the way he drives his car. Indeed, the horn of this green sportster does not honk obsessively when it comes to passing a car. Moravia explains it very well: for Torello it was a point of honor to pass everyone, but without honking his horn and without looking to see whether the way was clear, with his head down just like a bull. He had no need of

The Ferrari 250 SWB was produced in 200 copies.

The Ferrari 250 SWB expresses speed even while standing still.

a car to show off his arrogance; the money that filled his pocket, thanks to the thriving garage of which he was owner, made him quite arrogant enough.

Bruno Cortona, the protagonist of Dino Risi's 1962 film *Il Sorpasso* [The Easy Life], is quite another character. A rascal who has bungled everything away, including the people he loves, he has only one thing left: his white B24 Spider. This famous two-seater roadster, which in Italian is called a "spider," is the co-star of the film. Generous in performance, elegant in the design of the body, it makes up for all of Bruno's failures, his boorish behavior, and that one and always the same T-shirt and pair of pants he wears. Bruno, played by a great Vittorio Gassman, is convinced that, thanks to that car, everything is still possible, even playing with the life of an unknown student—Roberto, played by a very young Trintignant, who in the sun-drenched Rome of the August holidays hears the honking horn and comes to the window. The timid Roberto will follow Bruno in his exploits, hoping to get to Viareggio where Valeria, a girl he says he loves, is vacationing with her family. What's worse, Roberto himself will be the sponsor of this crazy race, paying the expenses with his savings, and in the end paying with his life. "You don't drink, you don't smoke, you can't even drive a car, what do you do for fun!" Bruno yells at him, pointing at the needle of the rev counter at the upper limit. "Don't look at the speedo, it goes much faster." Theater of action: the road, the Grosseto Aurelia; but, especially, the background emptiness, the artificial emphasis of the early years of the economic boom. The white, sleek B24 shoots into the bend, its horn blaring at the top of its voice. As Roberto Bessi writes in *L'auto e il cinema*: "At times the automobile has been 'personalized' to such an extent that, once it has absorbed the character and the desires of the human protagonist, it takes his place and overshadows him entirely. This is the case with the Spider in *Il Sorpasso*, whose impatient horn has entered common usage and describes better than any word the character of the protagonist." The movie technicians perform miracles to capture

Ferrari 250 SWB, 1960

When you drive it, you think all the other cars are stupid.

Paul Frère

The Ferrari 250 SWB is one of the models where the equilibrium of the body expresses better than others the theme of the high-performance sports car.

everywhere, in every moment, and in every way possible this dashing mechanical co-star.

If Bruno Cortona in *Il Sorpasso* is a significant figure in the story line of Risi's film, the streamlined body of the B24 anticipates themes that go far beyond that story. Pierre-Yves Thomas in his essay "Armonia innovatrice" [Innovative Harmony] remarks: "Pinin Farina was one of the champions of the rounded and enveloping style that distinguishes his production from the late-1930s until the mid-1950s. Once again the archetype may be considered the Cisitalia. While the world of automobile creation was evolving uniformly in this style, Battista Pininfarina in the early 1950s began to prepare a change of course, stretching the principal lines of his bodies a little at a time. A line is OK, he said, when it is taut and correctly bound up at the extremities. A car can be rounded without being soft. This is the key to the personality of the Lancia Aurelia B24 and the Alfa Giulietta Spiders." Pinin sowed the seed, in 1954, of a new way of understanding the body of a car. He feared the danger that forcing the forms to become rounded would make the line soft, and he recovered its equilibrium. The body of the B24 is tightly reined in, yet the tautness does not detract from but, on the contrary, exalts the plasticity of the design. And if to the eye this 2451 cc V6 seems already in motion when in fact it is standing still, this is due to the fact that its look weds perfectly with the designer's intention. And Pinin did an encore the following year with the Alfa Romeo Giulietta Spider. The new car's fortune is linked to its name—fresh, lively, quick off the tongue and quick off the mark, like its medium-power engine capable of the highest performance. Luca

"With this car you could go about your everyday business all week," Sergio Pininfarina recalls, "and win races at the track on Sundays."

following pages
The interior of the Ferrari 250 SWB attests to the car's elegance and refinement.

The flowing line of the Alfa Romeo Duetto highlights the scalloped flank and the cuttlebone-shaped tail.

Grandori writes in *Un sogno su quattro ruote*: "The Giulietta Spider filled a space that had always been empty, this side of the legendary Ferrari and Maserati grand touring cars. For the price of a tame Fiat 1400 you can drive a thoroughbred that does 100 mph easily, a pretty good speed for those days." Pinin himself will call it the "fair maiden," to distinguish it from the "fair ladies," famous and famed queens of the road. "The Giulietta Spider, which came out in 1955, was to become the premier means of picking up girls. The bachelor who tours the most fashionable sea-front promenades is the founding father of that extended progeny of strange animals, half-man half-machine, who do their courting with one foot on the accelerator and one on the brake. There wasn't much competition in those days, and for the most part

Alfa Romeo Duetto, 1966

A phase in the testing of the paint of the Alfa Romeo Duetto in the Pininfarina factory.

girls in the late 1950s have not yet assumed the aggressive, almost inhibitory bearing of their future descendants in miniskirts. On the contrary, they wore their skirts below the knee, sporting chaste two-pieces and a straw hat, to emulate the leading models: the diaphanous Grace Kelly and Audrey Hepburn. But this use of the car for picking up girls was still for the privileged few, and the Giulietta was a guarantee for someone with no trace of the charm of a Cary Grant—who at the wheel of an Alfa Spider broke so many hearts in *To Catch a Thief.*"

With the Giulietta Spider the industrial production of Pininfarina begins. Sergio explains: "This car had a fundamental significance for us—it meant the effective transition from craftsmanship to industrial processes. While the bodies produced until then were made on wooden molds edged with iron on which the sheet metal worker beat the panels, which were then welded and assembled, with the Giulietta Spider for the first time we used dies. Although this meant maximum investment, since the use of dies was much more expensive than the usual procedure, it also guaranteed maximum return, in the millemetrical precision of the parts produced. Once the dies of a hood or a fender had been cast, the presses made them perfectly identical each time. Sure, you need to produce enough units to justify your investment, otherwise… This, however, was certainly the case with the Giulietta Spider. We produced 30,000 of them over a period of ten years. We could never have dealt with an industrial line for the B24, of which we produced a very small number." Gaetano De Rosa writes in *Alfa Romeo Giulietta Spider*: "The word 'spider' is an English word meaning *ragno*. The name spider was originally applied to a type of small open cart, with disproportionately large, slender wheels. This antique definition, closely linked to the romantic age of the automobile, has now been lost. Who would have the courage to define, literally, as 'spiders' the convertible sports cars of today, so fast and brilliant, in which mechanical genius mixes with the inspired hand of the designer?… The 'spider' is a fascinating mix of speed, love of nature, unfailing vanity, and fun, all together in the country air. But above all the 'spider' is a symbol of eternal youth, which includes old memories and young emotions, and which is free of schemas and conventions… The 'spider' makes us one with nature, it gives us the possibility of admiring the scenery without the dull filter of sheet metal and glass. This car is a special pleasure, intoxicated by the air and the sun, perhaps covered with dust and with the windshield that's given a free ride to gnats… The

ALFA-ROMEO
MILANO
1300

Detail of the Alfa Romeo Duetto's cuttlebone-shaped tail, the symbol of an epoch.

The playbill of "The Graduate": Dustin Hoffman at the wheel of the Alfa Romeo Duetto during a break; a moment of the production: a technician waters the car, making rain.

Giulietta Spider became a symbol, the best-loved car in Italy in the period between the late 1950s and early 1960s.... The fair Giulietta has by no means gone out of style, neither does her success tend to diminish. For confirmation, just go see any show of classic cars and you will see for yourself the success obtained by the 'spider' of Portello today. Sure, it's not acclaimed like a Ferrari 250 California, or heralded like a Lancia Aurelia B24, but the Giulietta has a fascination all its own. To understand it, you have to drive it outside of the chaotic cities on a country road, cruising along at 30 miles an hour, admiring the scenery and chatting amiably with the lady at your side. Try it for yourself..."

In the story "Un amore" [A Love] by Dino Buzzati (1906-1972), a middle-aged architect from Milan falls in love with a would-be model he got to know in a brothel. They start seeing one another, and he joins her in Modena for a weekend. Then they drive back to Milan together in the red "spider" that the architect has borrowed from a friend. And so, getting onto the highway for Milan, "a beautiful sunny day, it was a quarter past three, driving a red open car, at my side a pretty and spicy and very modern girl who knows about everything that very modern girls need to know, and that's not all: at my side my loved one, in person, the most desirable woman of all women in the world, she obsession nightmare fatality mystery vice secrecy chic low-life big-city perdition, love, she at my side, a blue scarf with white polka dots knotted under her chin, provocative and haughty little country girl, going in an open car like this was fantastic, too bad there were no people around, there was no one to gauge her marvelous privilege of going on a May afternoon in a red roadster with a fellow like this at her side, little girl and not so

Some details of the Alfa Romeo Duetto, USA version.

The Alfa Romeo Duetto was produced, in various editions, for nineteen years.

A phase in the painting of the Alfa Romeo Duetto: water testing, 1974.

little girl, baby and woman, little flower and sin and all this was right there just take a look, oh! to go on like this and not to have to go to work, for the sun never to set and the road go on forever."

All over the world the Giulietta will be the standard-bearer of Italian style for roadsters. But what secret accounted for her role, and for her long life? Riccardo P. Felicioli, in his book *Sergio Pininfarina. Car Man*, writes: "In a certain sense it's the technical checkover that determines the aesthetic equilibrium and sobriety that distinguishes the entire Pininfarina production, that very special allure which gives all their cars a particular longevity—an extraordinary resistance to changes of fashion and taste." There will be no break in continuity when the founder decides to leave his post, to leave the field of battle. Felicioli will confirm this: "When Sergio Pininfarina was working with BMC on development of the design of several new models (A40, A55 and A105), the great designer Sir Alech Issigonis, creator of that extraordinary car, the Mini, after having watched him for a while with the suspicion and diffidence that, typically, a technician reserves for an artist, at the end concluded, reassured: 'But you are an engineer of style!'" Issigonis realized, seeing him work, that this young Italian wasn't pursuing form for its own sake, he was seeking a reconciliation of the technical and the aesthetic. "In fact it is precisely the love of technique, confidence in innovation, flair for experimentation, the attentiveness to everything that is advanced, modern, out of the ordinary, that will reveal to the careful and educated observer the *fil rouge* that links and connects in an uninterrupted flow the personality, intelligence and sensibility of the founder with the human and professional characteristics of his heirs." Sergio recalls: "The president of the Austin Morris, Sir Leonard Lord, invited us to Long Bridge to draw up the co-operation agreement. My father didn't speak English and immediately declared that he was putting me in charge of negotiating and sealing the deal. I got along fine with my prospective partners—I seemed just like a young Englishman myself at that time—and the president put me in charge of designing a small station wagon on the chassis of the A39. He had understood before anyone else that a family car, to be successful, had to offer plenty of usable space. So here I was with a chance to work with one of the greatest designers of all time, Alech Issigonis—later Sir Alech Issigonis. Together we made the Austin A40, a car that was light-years ahead of its competitors—it already anticipated future

The Fiat 124 Spider has a line characterized by the distinctive creaseline that runs along the top of the flanks, starting above the bumper. The line of the hood slopes down towards the front and concludes with a horizontal grill in the opening of the air intake.

hatchback cars. It was actually a sedan, but with the utility of a station wagon. In the meantime, on his own, Issigonis was working on the Mini, a car loaded with inventions that cannot even be repeated today. The Mini and the A40 left the shores of England to travel all the roads of Europe." The line of the body of the A40 clearly displays the continuity of intention of that criterion announced by Pinin, that "a line is OK when it is taut and correctly bound up at the extremities"; the body is no longer conceived as a smooth surface but rather as an architecture consisting of lines, planes and dihedrons. Pierre-Yves Thomas writes: "In spite of having created a style and presented a discovery that laid down the law, Pininfarina never stood still. Ever since the early 1960s it attempted to reconcile the enveloping style, made up of curves, with the great taut lines of the dihedral style, picking up the advantages and the less extremist aspects of both." From the rounded line, to the significance of the dihedron, to the reconciliation of the best of both—a design-criterion unremittingly in search of the archetype.

What, however, are the peculiarities that distinguished the culture of design in that period in a broader context? Bosoni and Nulli remark: "Beginning in the 1960s the culture of design, increasingly linked to the boom in consumer goods, took on an indisputably preeminent role, even within the vaster area of design in the broadest sense... Significant signs of the times were the progressive transition from architecture to design on the part of many architects and a consolidation of professional groups and associations, in a perspective of methodological and ideal unity that was progressively superseded by one of professional

Fiat 124 Spider, 1966

Preparing the advertising campaign for the Fiat 124 Spider.

The Italian director Mario Monicelli on the set of the film "Amici miei parte II" [My Friends, Part II], with a Spider Europa, a version of the Fiat 124 Spider, in 1982.

and competitive specialization. In fact the creative restlessness of the production of medium- and small-sized objects was matched by the frenetic professional and cultural organization of a variety of specialists (generally architects by training)—designers, graphic artists, and practitioners of the image in advertising, photography, and magazines—in concomitance with the rise of a new idea of consumption with laws of organization and a cultural influence all its own. The dawning Italian plastics industry with its surprising worldwide industrial development was to constitute one of the most significant poles of this period. A number of phenomena that were unusual for the Italian context in fact coincide in this sector. On the one hand the scientific research of Giulio Natta and his team at the Milan Polytechnic, with their discovery of polypropylene, attained a position of preeminence in the field, sanctioned by a Nobel prize in 1963. On the other, the phenomenon of Italian 'Good Design' was to attain a certain international renown in its own right, participating in the consecration of the Italian image abroad; it profited from the positive atmosphere of the sector, which found in the application of plastics a confirmation of those social values which many people were striving to make the basis of industrial design."

Soon enough, however, all this will reveal its negative side, as Joe Colombo pointedly remarks in his article "Antidesign," in *Casabella* no. 342 from 1969: "The nice, shiny, colorful product, whose formal aspect follows more or less faithfully the pure lines dictated by the laws of rationalism and functionalism that were established many years ago in those happy days of the Bauhaus and rationalist architecture,

For the American market
Pininfarina produces
the Europa Spider, which
is advertised and sold under
the Pininfarina name.

Fiat 124 Spider, 1966

The Fiat 124 Sport of Alcide Paganelli, Italian rally champion, in 1970.

not to mention Scandinavian furniture, not less enriched by pseudo-cultural substrates and by mannerisms of revival that return to the surface from time to time—these pure lines sing us the swan song of a design that creates isolated objects, now incinerated by the fashion of the day, and what's more we can witness the chaos that takes shape more every day in that microcosm which constitutes the human habitat, whose confused and disordered image is being reduced to a bazaar of wildly assorted useless objects, even if originally these objects were born of a designing that was rigorously rational and functional but isolated and uncoordinated." Still, Joe Colombo tries to find a way out of this drift: "If city-planning determined architecture, this in turn coordinated the interiors, which in their turn were to be filled with objects and elements of furniture whose design was not coordinated and certainly did not derive from the pure investigation of a problem that only in the best of cases could be dealt with in a field of simple applied investigation. The analysis of the problem should therefore work from the inside out, just the opposite of how it has normally worked until now. The space that surrounds a human being—naturally understood as a sample extrapolated from the collectivity, from the group—everything that constitutes this microcosm, in which a human being lives and

The Spider Europa, in Sergio Pininfarina's words is "a delightful car, from the formal point of view an even worthier daughter of the Giulietta Spider."

Fiat 124 Spider, 1966

A phase in the painting of the Ferrari 330 GTC.

The Ferrari 330 GTC participates in an Elegance Contest in Montecatini.

moves, is the field in which a series of investigations will have to be carried out that are intended to resolve the problematic of a new habitat, structured and coordinated, that is more in keeping with our times… It will without doubt be a new discourse that refuses *a priori* to take into consideration the individual elements that can construct a habitat, in order to create instead a unitary discourse."

Car design, however, is distinguished by the attention it pays to the moving object in its totality. The Giulietta Spider designed and mass-produced by Pininfarina for Alfa Romeo and the A40 produced by Austin Morris to a Pininfarina design are just a few examples of this intention. The automobile is the object that has to be studied in its complexity before it can be resolved in the aesthetic sense. It is not possible to start from aesthetic invention in order to resolve the complexity of the object. Only the consciousness of complexity suggests solutions of apparent simplicity. Thus a Ferrari has to be in the first place a Ferrari, fully expressing the culture of the Maranello sports car manufacturer; it cannot wear clothing that is not its own, leaving it to the tailor to exalt its identity. Among the projects newly undertaken by young Sergio Pininfarina, the relationship with Maranello grows in the respect of the complexity of Ferrari design in its most faithful and exhilarating interpretation. The Ferrari 250 SWB produced in 1960 is one of the most significant examples. Roberto Segoni in his essay *La carrozzeria Pininfarina vista da…* [Pininfarina Bodywork Seen by…] writes: "A car without the slightest flaw, but it is also one of the examples where the equilibrium of the body expresses better than others the theme of the

The Ferrari 330 GTC presents a flank lightened by a creaseline along its length and an ogival air intake in the front. Starting in 1966, six hundred of them were produced.

The Ferrari 330 GTC on display at the 1967 Brussels Motor Show: Pininfarina takes great pains over extreme simplicity of line and absolute compactness.

high-performance sports car and attains the highest expression in an absolute sense. Not only is there nothing to criticize about the 250 SWB—this car demands interpretation, it should be made the object of investigation of the scientific study of signs. It demands, that is, a semiological analysis of the specific elements, valencies, relations and forces that the various elements of the modeling possess in expressing a form that is sleek, powerful, aggressive, imposing in the importance of the car but not imposing in size, as was the case with the old Isotta Fraschini. This car is imposing because of its sense of power, its intrinsic dynamics, even in the potential state; that is, it expresses speed even while standing still; its form is dynamic even when it is at rest." The cause of the formal equilibrium of an object, especially of a complex object like the automobile, cannot be so easily reduced to an inventory of signs. Otherwise, once the signs are distinguished, the complexity can be resolved in the formulation of a simple recipe. The 250 SWB, as a technical-aesthetic synthesis that was attained at a precise moment of a process of evolution, of Ferrari in particular, and of Pininfarina, constitutes itself as an archetype. The matter can be repeated but not the form. Two hundred examples of the Pininfarina designed 2953 cc, V12-engined 250 SWB (short wheelbase) will be built by Scaglietti in Modena. Sergio recalls: "With this car you could go about your everyday business all week and win races at the track on Sundays. Paul Frère, a journalist and a former Formula One driver, once said: 'When you drive it, you think all the other cars are stupid.'"

But Ferraris are the dream of the many and the need of the few. Most people satisfy their need for a car with a purchase in keeping with their means. And in fact economy cars are the best sellers: the Fiat 500 designed by Dante Giacosa will win the coveted Compasso D'Oro prize in 1959. And on the domestic front—regarding, that is, questions of the home—what will be the fate of the new furnishings, of the objects of furniture design? They are neither the dream of the many nor the need of the few—in the home we still find a marked preference for time-honored presences. It is curious to observe how not even the ecological revolution expressed by the new forms of home furnishing was well received; people continued to prefer their imitation baroque dust catchers, even the sectionals. The products of design, then, as far as the home is concerned, were reserved for a precise and particularly restricted cultural milieu, which was not determined by the bank accounts of its members. In other words, it could by no means be taken for granted that people who could buy a Ferrari 250 SWB were willing to furnish their homes with furniture of analogously advanced conception—but the same held true for buyers of the Fiat 500. So, within that restricted cultural milieu, as Bosoni and Nulli point out: "If until the early 1960s Scandinavian taste had dominated, around that time an Italian model began to take hold. Together with the new myths created by the

Ferrari

The Ferrari 330 GTC is described as being easy to handle, quick off the mark, easy to drive, practically adorable; the details shown here of the bodywork and the instrument panel cannot but confirm this judgment.

In the course of production, the headlights undergo an important modification: from faired-in to retractable, for the sake of greater efficiency.

economic and social development underway in Italy, the question of furnishings became a very popular theme, as is shown by the birth of several magazines—*Interni* [Interiors] in 1954, *Il Mobile Italiano* [Italian Furniture] in 1958—and fairs and exhibitions specialized in the sector, such as *Le Selettive* [The Selectives] of Cantù in 1955, the *Biennale dello Standard* [Standard Biennial] in 1958, and the Furniture Show in 1961. But if on the one hand this meant an affirmation of modern furnishing that was attracting the attention of furniture manufacturers, on the other there was the spread of a certain conformity and mannerism that, apart from people of heightened sensibility, expanded rapidly, distorting the premises of the rationalistic root and interpreting all previous research projects in an exclusively formal sense."

It was Dino Gavina, one of the most sensitive furniture producers of the time, who became a leading advocate of the need to stem the formalist tide. Taking stock of the rationalist experience, he decided to oppose the products manifesting the crisis of the day with the tubular furniture that Marcel Breuer, the grand master of the Bauhaus, had designed in 1929 with an eye to large-scale mass production. Claudio Cerritelli comments in his essay *Dino Gavina*: "Having made his indisputable point, Dino Gavina then proposed the *Ultrarazionale* [Ultrarational], which unhinged techniques of the coupling of materials and formal principles that had been precluded by the dominance of rationalist dogma and that now cry out for vengeance—not, however, without maintaining certain ties with the culture it had rejected. Then, with *Ultramobile* [Ultrafurniture], Gavina tightened his relationship with art, not only because he drew directly upon original and

The Ferrari 365 GTB 4 Daytona, from 1968, is unquestionably one of the highest aesthetic expressions of car design at that time. Its name evokes the races won by Ferrari on the legendary Daytona circuit.

eccentric forms he found in Man Ray, Magritte, Matta, Warhol, etc., but also for the irony and the sarcasm that he primed on the predictable terrain of design, transforming it into an art of the imaginary, a body of thought that upsets the usual logics, a biological impulse that orients any type of choice." This continual search for the archetype made Gavina not only the guiding light of Italian furniture design, but one of the most important figures in the history of design in this century. "To Gavina's mind, the problem was that in the second half of the 1950s the figure of the designer still had not become strong and consolidated in Italy; the word 'design' had a fragile and hesitant identity. He realized that to design a piece of furniture an injection of artistic culture was necessary, a nutrient that came from the contemporary avant-garde, precisely in its guise of a tradition of the new—from futurism to surrealism, from dadaism to pop art, from action painting to programmed art. What attracted him was the lesson of the decontextualization and the appropriation of objects of common use through a new function. This was the mainspring of various intuitions that permitted him to see the things around us in a different way—the forms that shape everyday living, the function of use linked to aesthetic pleasure. His relationship with artists was intense and surprising, from Marcel Duchamp to Man Ray, from Alberto Giacometti to Sebastian Matta, from Ettore Colla to Umberto Milani, from Julio le Parc to Allen Jones—a mesh of ties that never became empty patronage but was true fellowship, an osmosis between intelligences who affirmed thought itself, not just its dogmas. Indeed, wasn't it Lucio Fontana in 1954 who

In 1971 the Ferrari Daytona is also exhibited at the Milan "Salone delle Arti Domestiche," as is shown by this document of the time.

With the Daytona Pininfarina abandons the classical elliptic form of the air intake, which had been a distinguishing mark of the Ferrari until then, in favor of a linear design.

brought to Gavina's attention several young architects, including Pier Giacomo Castiglioni, on the occasion of the tenth edition of the Triennale? Wasn't it Fontana himself who suggested he put these new talents to work?"

Marco Zanuso, with the collaboration of Richard Sapper, designed the Lambda chair for the Gavina company, in sheet metal painted in strong colors to a high gloss; it entered production between 1962 and 1963. François Burkhardt explains: "The idea of the Lambda came from the technique of the car-body beaten by hand by master coachbuilders. Zanuso had taken an interest in this technique in the 1950s when he visited the Pininfarina factory in Turin; he had been impressed with the workers' skill in beating the sheet metal on wooden bucks, to shape the spoilers of a custom-built car. He later decided to apply this technique to chairs manufactured in a limited number of copies, thus giving rise to yet another innovation in the furniture field." If this confirms the impossibility of employing dies and presses without a sufficient volume of production to justify their use, it also confirms the scant interest of the market when it came to materials used in unusual ways for the construction of objects of furnishing. But Zanuso's innovations, of course, did not stop there. "With the Grillo telephone Zanuso designed for the first time in telephonic history a version of a small portable phone, with integrated earphone and dial that completely changed the phone's appearance, in response to the producer's request for a one-piece telephone. Grillo is the object that revolutionized sixty years of practice in the conception, construction and use of the telephone, creating a wholly new type. The research had been carried out with the idea of transforming certain habits and of joining together the standard elements of the phone (an idea that was applied constantly to Zanuso's typologically innovative objects). The aim was to abolish the division between the base and the earphone and to make the object lighter and smaller, in order to arrive at a telephone conceived as a compact object that one picks up, brings to the ear, and then puts back down, and that opens and closes like a shell to keep out the dust when it's not in use. It was a revolutionary telephone that blazed the way to a whole series of wireless and portable phones, all based on the same principle."

And then Zanuso turned his attention to radio and television. "The Doney and Algol II televisions produced in 1962 were based on the principle of the shunts of transistor television (Doney was the first of this kind in Europe). Once the necessary elements had been obtained,

Ferrari 365 GTB 4 Daytona, 1968

The cleanest design I've ever seen.

Bob Blake

The Daytona was the best car possible, as long as Enzo Ferrari insisted that the horses had to stay up front.

one proceeded to the synthesis of the appliance container, which presents a structure that is different from other TV sets. Here, the components are grouped together on the basis of their operative function as elements of different shape: the tube, the group of amplifiers with the control circuits and volume circuits, the feeding group with its circuits, batteries, transformers, controls and channels, and the internal feeding group with the battery and the internal circuits. A structural element is assigned to each of these functional elements that makes it possible to connect the four parts, spatially defining these components and their frame in order to place them all in the envelope that covers them. The procedure is a sort of piece-by-piece assembly, which will make it much easier to get to all the parts in case of repair. This way of organizing

Ferrari 365 GTB 4 Daytona, 1968

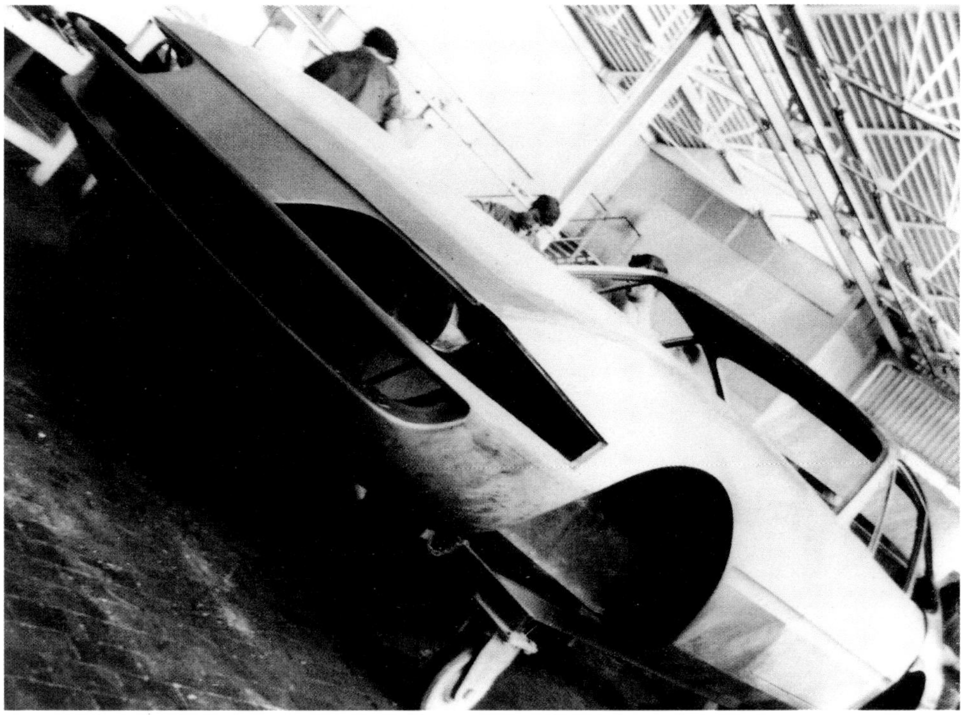

A phase of the bodywork at the research center in Grugliasco.

electronic parts was absolutely new, making the appliance an experimental object of a new type." And if the radio, and especially the television, had been seen until then as a totem to be installed in the home in a place of honor, with the entire community kneeling down before it, Zanuso revolutionized its meaning. For the first time the TV set is a piece of furniture, part of the decor, a participant in the domestic landscape. Brionvega was operating in a market that is one of the most sensitive to the growth of demand, and Zanuso's project satisfied at that moment its qualitative discrimination. They even went on to produce a model for the renewal of the entire productive sector. And so, around that time, the television set began to make history, while chairs had to cope with an age-old tradition. So the Lambda remained in the ambit of the experimentation of techniques and materials, while Doney and Algol II were mass-produced, and in a big way.

In the very years that saw Sergio take his father's place as president of the company, Pininfarina confirmed the possibility of mass-producing objects of high formal quality. In 1966 the "spiders": the Alfa Romeo Duetto and the Fiat 124 appeared in that market niche which had already seen the undisputed success of the Giulietta Spider. Riccardo P. Felicioli notes: "Beginning in 1966, the Alfa Romeo Duetto and the Fiat 124 Spider, with a production that shot up to some hundreds of thousands of cars (all together, over 200,000 for the 124 and 140,000 for the Duetto), won for Pininfarina a truly industrial dimension, irrevocably classifying it as a mass-producer, even if on a limited scale." Limited, however, only if compared with the number of vehicles produced by the big car-makers. But wasn't it Sergio's objective to produce precisely those models that would never have justified the investments of the big industries? "In that same year Sergio decided to give the company a more rational organization, defining the study and research department more sharply, and transferring it to an area on the other side of the street from the production and assembly lines. At last the two souls of the company were made visibly distinct. And in 1966 the decision was made to build a wind tunnel, which will be inaugurated in 1972. This was the occasion in which Sergio Pininfarina showed his extraordinary technical and entrepreneurial farsightedness. In the 1960s, as an effect of the low fuel prices, the great theme of aerodynamics, which in the 1930s had aroused such great passion and enthusiasm, decisively influencing car design, was effectively pushed out of the scene. It retained its importance, obviously, where motor-racing was concerned, where a better drag coefficient meant greater speed... In the mid-1970s, following the Yom Kippur war, oil prices went through the roof, and we were faced with the irreversible and terrifying possibility of a permanent shortage of energy resources. All of a sudden, in the research centers and design departments of the automotive industry all over the world, the question of

Ferrari 365 GTB 4 Daytona, 1968

The 504 Coupé and Cabriolet is the first car that Pininfarina constructs for Peugeot.

aerodynamics moved right back onto center stage. Pininfarina, thanks to the felicitous intuition of its young president, had at its disposal the instrument that makes it possible to scientifically measure the relation between the form of that geometrically complex object which is the automobile and its capacity for aerodynamic penetration; it was in a position to offer its clients a service that had become crucial in the economy of the design process."

Luca Grandori, in *Un sogno su quattro ruote*, evokes that climate which gave rise to the Alfa Romeo Duetto: "American television has become a perpetual motion kaleidoscope that functions around the clock. Only once did the TV screen go blank—one night in September of 1965 when a blackout left the entire city of New York in the dark. The blackout demonstrated the vulnerability of the ultra-automated metropolis, but the greatest concern of American technicians and scientists was focused on the levels of air pollution; 'greenhouse effect' had not yet entered the language. But the United States, always more farsighted than the rest of the world, gave itself the first law regulating automobile exhaust emissions. This anti-smog regulation, however, could not intimidate Ford Motor's designers, who promptly launched the high-powered Mustang. To underline its merits Ford sent a whole team of cars to compete in the Tour de France, and the Mustangs, which were named after the most famous American fighter-plane of World War II, swept their adversaries away in the 3000 cc category. With the Mustang the US attempted to make it in the European auto market, but without much success. The Italian car buyer found these American cars heavy, unwieldy, exuberantly powerful, but only thanks to colossal cubic capacity. The merits of Italian cars, with Alfa Romeo at the fore, consisted instead in that formula which included lightness, power and stability. These qualities were to be found, and better than ever, in the Duetto 'spider,' with the amply tested 1570 cc 4-cylinder engine and the distinctive cuttlebone-shaped tail. Its official consecration took place in the film *The Graduate* with Dustin Hoffman; the only other equally famous movie 'vehicle' of the day was Dennis Hopper and Peter Fonda's Harley Davidson chopper in *Easy Rider*. The Duetto completed, with the GT coupé and the Giulia Berlina (which in the meantime had been updated to the Super version), Alfa's winning trio in the second half of the 1960s."

"Duetto" is a rather strange name for this cuttlebone-shaped "spider." The term is a lot

A postcard of the early 1900s: Avenue des Champs-Elysées, Paris; on the right, one of the first automobiles.

Peugeot 504 Coupé and Cabriolet, 1969

The top of the Cabriolet Peugeot 504 folds up completely.

The comfortable driver's seat of the Peugeot 504.

closer to the world of opera, where plethoric tenors and sopranos well-set on generous and statuesque hips weave their elaborate melodies, than it is to that of the road, where the winning qualities of the Duetto are its lively spirit and sprint. An Alfa, the 1900, was James Dean's first car—the actor who symbolized the restlessness of youth in post-war America. After defying death in fiction, driving an old Ford in Nicholas Ray's *Rebel Without a Cause* in 1955, James Dean encountered death in reality, on the California straightaways, defying speed limits and big police motorcycles, at the wheel of a Porsche. Another Alfa, the Spider Duetto, will also be Dustin Hoffman's first car—as Benjamin in Mike Nichols 1967 film *The Graduate*, with Anne Bancroft, Katharine Ross, William Daniels, Elizabeth Wilson, based on the novel of the same name by Charles Webb. The film is a merciless portrait of the conventional, well-bred and "respectable" America of the 1960s, where the hero paints us quite a convincing picture of the restlessness of his generation. A felicitous choice for the role, Hoffman was not yet known to the general public as the great actor he was to become. Upon his return from college where he finished at the top of his class, Benjamin Braddock is received by his father—especially by his father—as a sort of trophy to be shown off to his friends. They all want to touch him, embrace him—and to touch and try out the red "spider" he received as a graduation present from his parents. This is just one example, but the young man feels stifled by the atmosphere, and he shows his impatience openly. What's worse is that he has no way out; or rather, coming home after the university, he cannot find the willpower to make up his own mind about what to do with his life. He even allows himself to be seduced by Mrs. Robinson, the dissatisfied wife of his father's business partner. But when he re-encounters Helene, the Robinson's daughter, he falls madly in love. Both of them understand they have nothing to do with that world which conditions them and steers their actions, and Benjamin doesn't want to lose the chance to free himself and Helene. He will manage to snatch her away from a future husband, right before the altar, in one of the most famous scenes in film history. They will escape from the furious disappointment of that world which treated them both as trophies to be put on display. And the little red "spider," cuttlebone-shaped, that Hoffman drives in the film with good temperament and considerable nonchalance, will be the great novelty on the American market, opposing its freshness to the traditional unwieldy presence of the powerful Mustangs.

The Peugeot Cabriolet 504 has a good balance of volumes and trim lines.

The hood of the Peugeot 504 slopes down towards the front and twin headlights flank the radiator air intake, which is surmounted by Peugeot's lion symbol.

"Remaining in the sphere of the society news of the period," Giancenzo Madaro notes in *Alfa Romeo Duetto*, "at the presentation of the Spider 1600 in New York, with the participation of about 1300 personalities of the world of industry, politics, art, culture and fashion, the soprano Anna Moffo drew the lots for a Duetto that was won by the American fashion model Heidi Gover." It's perfectly true that Anna Moffo was from the opera, but this soprano had the body of a model; indeed, she was so far from the stereotype of her category that she easily could have sung duets in the Alfa "spider." "They were the years," Luca Grandori recalls, "of Saint Laurent clothing inspired by Mondrian paintings, of the celebrated ateliers of Balenciaga and Givenchy in Paris, and of the success of Emilio Pucci—once Mussolini's pilot, the Florentine marquis was now designing high-style dresses, pants, blouses and even bathing suits." So the fashion world too took part in the American presentation of the Duetto; indeed, one of its vestals, Heidi Gover, won the car, with the central scallop running the length of the side, its headlights faired in plexiglas, and two air intakes on either side of the radiator grill, delimited by the partially wraparound bumper. "The Alfa Romeo Duetto is the daughter of an unbeatable car," Sergio notes. "But which one of the two, thirty years later, can still be considered worthy of consideration today, from a formal point of view? Perhaps the Giulietta, it's purer. We continued to produce the Duetto, which was a little more showy, in various editions, for fully nineteen years."

If the Giulietta Spider moved Pininfarina from the craftsmanship to the industrial dimension, and the Duetto manifested its potential, it was the Fiat 124 Spider that truly exalted the productive capability of the company. The car, with a double overhead camshaft engine, has a line characterized by the distinctive creaseline that runs along the top of the flanks, starting above the bumper and extending to the tail, which is truncated in the form of a broad V. The line of the hood slopes down towards the front and concludes with a horizontal grill in the opening of the air intake. Sergio recalls: "It is the car that without doubt we produced in the greatest number—more than 200,000. For the American market we even came up with the Europa Spider, which was advertised and sold under the Pininfarina name by itself. A delightful car from the formal point of view, an even worthier daughter of the Giulietta Spider. The rally version was also very good—big tires, automatic seats, pretty rigid but an exciting car."

504

The Ferrari BB controls its power very well and upholds criteria of lightness based on the study of construction materials.

It was a magical epoch, inaugurated in 1955 by the Giulietta Spider, carried forward by the Duetto of 1966, and concluded with a rush by the 124 Spider of that same year. "Swiftness, swiftness, joyous victory over the sad weight, fever air, thirst for wind and for splendor multiplied by spirit in the bony mass. Swiftness, first-born of the stretched bow called life." Who knows in which of the three—strictly red—cars this singer of swiftness would have put the two lovers of *Forse che sì, forse che no* [Perhaps Yes, Perhaps No], caught up in their frenzied race? In 1910 Gabriele D'Annunzio could only avail himself of his fervid imagination, for speed was still an emotion to be conquered. The poet was living at the Capponcina in the Siena hills and possessed a red Florentia, whose top speed was only 45 miles per hour. Vittorio Pirlo, in his essay *D'Annunzio e l'automobile*, writes: "In the Poet the exaltation of speed is a conception of life and, therefore, the order to the driver is one and one alone: *go fast*. A passion for which D'Annunzio had been considered dangerous ever since the days of the Florentia. If he was stopped, his justification was that, if he were not inclined to break the speed limit, his name would no longer be Gabriele D'Annunzio. Hunted by policemen who slapped him with one fine after another, he objected that they wanted not only to do away with speed but with the automobile itself… His irresistible yearning for speed and for the infinite could never get enough roads and space, and this anguished limitation led him to fight relentlessly for newer and better roads; not the last of his battles was the one in favor of the Gardesana Occidentale, which he will call Neandro. In this connection it's curious to note that in his library there were publications regarding the treatment of roads with new bituminous applications. This, too, was an aspect of his interest in the automobile and in the many problems that derive from it." And while Marinetti was exalting the unbridled and raging mechanical monster, D'Annunzio was celebrating the form in motion, comparing it to feminine beauty. This is how he thanked Giovanni Agnelli, the founder of Fiat, for his gift of a 509 Cabriolet: "My dear Senator, in this moment I return from the field of Desenzano with your car that, it seems to me, resolves the question of gender we have discussed before. The automobile is feminine. This one has the grace, the sleekness, the liveliness of a seductress, but also a woman's easy lightness in overcoming all roughness."

If the Fiat 509 had its poet in Gabriele D'Annunzio, some forty years later thousands of voices will be singing the praises of the Fiat 124

Ferrari BB, 1971

The Berlinetta Boxer is a direct descendant of the research prototype Dino Speciale of 1965.

Spider, enchanted by the grace of its form and the liveliness of its temperament. But also for the automobile, which is, still and all, only a machine, any form of recognition—be it Marinetti's masculine aggressiveness, or D'Annunzio's feminine sweetness—confirms the curious habit of no longer seeing machines as objects but rather as living beings, "in possession of all their faculties." This is a habit that car owners little by little have made their own. It goes back, however, to the beginnings of mechanization, about two centuries ago, when man broached the use of mechanical prostheses. Carlo Emilio Gadda (1893-1973) gives us his view on the matter in the essay "L'uomo e la macchina" [Man and Machine] in his volume *Le meraviglie d'Italia* [The Marvels of Italy]: "The machine, which in the last couple of centuries has played a greater and greater role in the history of men, and has mechanized their civilization (to some detriment, perhaps, of the reading of the Gospels), has also uncorked a whole new animistic phraseology, which would see the machine as participating in our feeling and thinking and doing and discerning. The workman himself sometimes speaks of the machine as of a thinking creature. He praises or disparages it as it deserves, hates or loves it, like a poisonous or a useful beast. In fact, the machine is nothing other than the carrying out of a mental procedure of ours, externalized and automatized in a practice. Under the direction of our thought, the machine substitutes for our movement and our muscular exertion—moved (today) by an energy that is external to the biological world."

Two fins reach back to the truncated tail of the BB and contain the spoiler designed to reduce aerodynamic resistance.

Ferrari BB, 1971

The horizontal louvers of the radiator air intake, placed in the front, are directly incorporated into the bumper.

Sergio Pininfarina and Bill Mitchell, vice-president of General Motors, at Grugliasco, in front of the wind tunnel, during the tests with woolen yarn in 1972.

In the years in which, as Guido Piovene put it, "specials were being mass-produced", the soul of the Ferrari range was the 3967 cc V12-cylinder 330 GTC coupé. This car, of which about six hundred were produced starting in 1966, did not present aesthetic solutions of great note—the flank lightened by a creaseline along its length, the ogival air intake in the front. Sergio explains: "I'm especially fond of this car, since I drove it for a long time myself. We took great pains over the extreme simplicity of the line, the absolute compactness. Easy to handle, quick off the mark, easy to drive—it was one of the most adorable Ferraris I ever experienced." And the Daytona, the last thoroughbred front-engined Ferrari, must have had a truly generous soul. Piero Casucci, author of *Jaguar, il fascino di un nome* [Jaguar, The Fascination of a Name], notes: "It is significant that Bob Blake's personal car is a Ferrari Daytona, "The cleanest design I've ever seen," he says. And the Daytona is Blake's third Ferrari, he's had two 330 GTCs previously." In calling the Daytona the cleanest line he'd ever seen, Blake was calling the Daytona the highest aesthetic expression of car design at that time. And Blake knew what he was talking about, because Jaguar could boast such beautiful cars as the D- and the E-Types, classical examples of English style. "The Ferrari 365 GTB 4 Daytona was truly an important car," Sergio confirms. "In its name, Daytona evokes the races won by Ferrari on the legendary Daytona circuit. The car dates from 1968, two years after Lamborghini had launched the Miura, with its centrally-mounted engine. The Daytona is a good six inches higher off the ground than the Miura, due to the underbody space occupied by the exhaust pipes. Despite this unquestionable handicap, we managed to give

Assembly line of the Fiat 130 Coupé.

Automobiles are driven on roads all over the world, bearing witness to the excellence of the Italian line. What, however, in the 1970s, is the testimony of the large family of Italian designer-objects? "There is no doubt," Bosoni and Nulli assert, "that on an international level Italian design of the 1970s was distinguished by a sort of natural elegance, which nonetheless appeared to depend more upon an empirical solution of the problem than upon an appeal to theoretical principles. What this design did show was a flair, a freedom in the formal touch, that was in close accord with the progressive change in what the elements that make up the environment meant—elements that were progressively required to be less severe and more escapist. This also describes very well the transition of the aesthetic principles of the modern from its minority condition of avant-garde, of rupture, of the forced advocacy of change, to a conception that sees as its objective and as its instrument the brilliancy of the solution, the ability in reconciliation, the will to a coincidence with what exists. The main characteristic of the design culture of the 1970s was thus in the first place the constitution of a double line (in some cases of a double consciousness, since we are dealing with a single person) in the designing itself. On the one hand there was the line linked to the world of industrial production that followed the clear-cut leap towards specialization and the sophistication of methods and controls. On this link there also weighed, in cases of the highest professionalism, both a doubt about the authentic social necessity of one's operations and the drop in genuine ideal tension apart from the tension of one's own private creativity. On the other, there was the world of poetic license,

the car plenty of sprint, concentrating on the length, on the taut lines, on the scallop along the flanks. As regards the front, apart from the headlights that underwent an important modification in the course of production—from faired-in to retractable, for the sake of greater efficiency—we abandoned the classical elliptic form of the air intake, which had been a distinguishing mark of Ferraris until then, in favor of a linear design." With the 4390 cc, V12-cylinder, 365 GTB Berlinetta, Pininfarina reveals the limits of front-engined high-performance cars. The Daytona had in fact been the best car possible, as long as Enzo Ferrari insisted that the horses had to stay up front. It thus remains a cornerstone of experimentation, awaiting the day when sophisticated technology will make it possible to put the fiery colts in harness once more.

The success of the Pininfarina line is attested too by the many awards that were received over the years.

The Fiat 130 Coupé is a car of extraordinary purity, marking the peak of design based on taut lines and plane surfaces.

Sergio Pininfarina

of playfulness, which keeps an eye on its relations with the visual arts, with fashion, with fast changes in taste. Often we had the same design groups working in both camps, which somehow compensated one another. The problem that was to come out later on was that this dichotomy, taken at the beginning for a positive transfusion of creativity by one side and of technique by the other, between craftwork and industrial investigations, generally tended to fall into a redundant multiplication of forms and of things, to the sole benefit of the system of commodities that was linking up more and more closely with the system of the image."

And I, in my story "Le donne i cavallier l'arme gli amori..." [Women cavaliers arms loves...] in *Stileindustria* no. 8, December 1996, wrote: "The design of that period, at least in the furnishings sector, was unable to get beyond form; it did not pursue the justification of its project; it did not resolve the necessary technological advance. The result was the elimination of most of the companies that had advocated it." In fact the high fashion of furnishings was quite distinct from prêt-à-porter: mass-produced unique designs. "It was not by chance," Bosoni and Nulli emphasize, "that those very years saw the great success of Italian fashion with the prêt-à-porter phenomenon in Milan, which was an ideal location thanks to its tight links with industry and commerce, and to its strategic position in the European and international market. This, in any case, was a phenomenon that expressed, in the best of cases, a genuine design approach for the production processes of the clothing sector." Thus furnishings design came to surrender the palm—in Milan, its place of origin—to fashion design.

Sergio Pininfarina and Renzo Carli in front of the wind tunnel.

The Fiat 130 Coupé is in every respect a sedan, with the beauty and the lightness of line of a large coupé.

Fiat 130 Coupé, 1971

testarossa

The very name Testarossa evokes racing myth.

The mythical Testarossa, racing in 1959.

But the 1970s also marked the debut of the informatics era. Designed by Steve Jobs and Steve Wozniak, the Apple II, of 1977, may be considered the first personal computer intended for the general public. This computer fitted right on your desk, while the Electronic Numerical Integrator and Calculator, or ENIAC, completed in 1946, occupied 160 square meters of space and weighed in at some 30 tons. Inventions such as the transistor in 1947 and the integrated circuit in 1959 made it possible, progressively, to reduce the size of the hardware and create more powerful artificial intelligence, the software. And so Marcello Nizzoli's Lettera 22 of 1960 and Ettore Sottsass's Valentina of 1969, symbols of that object called the typewriter, will soon be superseded by portable computers. The mechanical prostheses lose out in the end to the virtual.

Pininfarina was to take advantage of the new technological advance to create, in the late 1960s, "a center for calculation and design, called the DEA, equipped with the Touching Machines," notes Riccardo P. Felicioli, "a decisive step towards that process of the automation of design which will be fully developed in the following decades." In 1972 at Grugliasco Pininfarina inaugurated its wind tunnel. The technological revolution, far from impugning the firm's basic principles, reinforced its conception of design, which was refined in avant-garde conditions of testing and experimentation. "Pinin's defining gaze, his ability to sculpt surfaces and spaces with his own hands, was replaced by a more broadly coordinated gaze, by the examination and dialogue that is possible in teamwork. And this was possible because Pinin's aesthetic vision, his

Tests in the wind tunnel.

Ferrari Testarossa, 1984

Sergio Pininfarina accepting the award for the design of the Ferrari Testarossa at the Los Angeles Auto Expo in May 1985.

Inauguration of the 1984 Paris Salon: world preview presentation of the Ferrari Testarossa. Note the large air intakes, which reinforce the car's powerful and aggressive image.

professional authoritativeness, his experience and his knowledge, had been interiorized by his most direct successors and then, right down the line, by everyone who had taken part, through the years, in the group of Pininfarina designers. In this way a culture of design was formed and consolidated, shared by all concerned, and sufficiently stable. This stylistic continuity is thus the consequence of a particular approach to design. Sergio Pininfarina called it a way of expressing creativity that is not only an expression of individual talent, tied to the sensibility and the aesthetic intelligence of the single designer, of the artist who invents new forms and discovers beauty. Pininfarina managed to systematize its experience of car design—to generate a culture, a body of knowledge and of skill, an aesthetic vision that is reflected in all its products—in a way that was relatively independent of the personal characteristics of each designer."

Precisely this culture of design, this body of knowledge and skill, this aesthetic vision, is what will be reflected in the design and construction of the Peugeot 504 Coupé and Cabriolet of 1969. There had been relations between Pininfarina and Peugeot ever since 1951. With the model 403 of 1953 Peugeot had sold more than a million cars, an extraordinary number for that time, but "the 504 Coupé and Cabriolet was the first car Pininfarina actually constructed for Peugeot. A dream car," Sergio recalls, "but the people who dreamed most of all were the French, who had been so accustomed to the rather tame lines of the 403 and the 404. The car was a sort of free "tone-production"—a free "touch," to use the musical term—as regards the theme of the sedan, which we had designed, that won the title of Car of the Year. The Coupé and

Italian style is not a style;
it is a state of grace.

Mario Bellini

The Testarossa is absolutely Italian, in the state of grace of Pininfarina.

Cabriolet have a good balance of volumes, trim lines. The hood slopes down towards the front. Twin headlights flank the radiator air intake with its three horizontal rods, surmounted by Peugeot's lion symbol. The Cabriolet's top folds down completely. The roof of the Coupé is made considerably lighter by an ample greenhouse and the wraparound windshield gives maximum visibility."

James Joyce (1882-1941) published *Dubliners* in 1914. "Each blue car, therefore, received a double measure of welcome as it topped the crest of the hill, and each cheer of welcome was acknowledged with smiles and nods by those in the car. In one of these trimly built cars was a party of four young men whose spirits seemed to be at present well above the level of successful Gallicism: in fact, these four

Ferrari Testarossa, 1984

young men were almost hilarious. They were Charles Ségouin, the owner of the car; André Rivière, a young electrician of Canadian birth; a huge Hungarian named Villona and a neatly groomed young man named Doyle. Ségouin was in good humour because he had unexpectedly received some orders in advance (he was about to start a motor establishment in Paris) and Rivière was in good humour because he was to be appointed manager of the establishment; these two young men (who were cousins) were also in good humour because of the success of the French cars. Villona was in good humour because he had had a very satisfactory luncheon; and, besides, he was an optimist by nature." This cheerful crew is going by in a splendid car, open and blue. In this story, "After the Race," Joyce was referring to facts that had taken place in the decade preceding 1914. It is curious to note how, in 1969, almost seventy years after the events of the story, Peugeot's motto, "Peugeot est sérieux," contrasts with the lively color of the French cars at the beginning of the century. But the 1769 cc, 4-cylinder, 504 Coupé and Cabriolet, designed and produced by Pininfarina for Peugeot at the end of the century, are far closer to the splendid blue cars that enchanted the Dubliners.

But niche cars were attracting more and more international customers. When in 1968, with the Daytona, Pininfarina revealed the limits of front-engined high-performance cars, traction-control technologies had not yet made it possible to overcome those limits. And so Enzo Ferrari made up his mind to produce his first mid-engined thoroughbred: the 4390 cc, 12-cylinder, Berlinetta Boxer, a direct descendant of the research prototype, the Speciale of 1965. About a thousand cars were constructed. "A

Ferrari Testarossa, 1984

history-making car," Sergio notes. "It marked Enzo Ferrari's conversion to the most justified and most justifiable technical solution. We designed and built the BB with great enthusiasm: low line, flowing, a groove along the flanks ideally divides the body into two parts, so that they appear superimposed. The horizontal louvers of the radiator air intake, placed in the front, are directly incorporated into the bumper. On either side, there are the intakes for cooling the brakes. The headlights are totally retractable and they open and close by electronic control. The directional indicators are incorporated in the line of the car, right at surface level. The sharp rake of the windshield fits and joins the roof to the hood. Two fins reach back to the truncated tail and contain the spoiler designed to reduce aerodynamic resistance. The engine cover incorporates carburetor air intake fairings. The car controls its power very well and upholds criteria of lightness based on the study of construction materials: chassis and body in steel; hood, trunk and doors in aluminum; lower, front and rear shells in reinforced resin. On the louvers of the front air intake, rightfully proud, stands the Prancing Horse."

While in 1969 the Peugeot 504 Coupé and Cabriolet made the French dream (and not only the French), in 1971 the Fiat 130 Coupé did not attain its due success on the Italian and international market. The 504s strengthened Peugeot's credibility, even though they were confined to a rather tight niche; but the 130 Coupé won such little favor with Fiat customers that its splendid line was largely overlooked. "It is a car of extraordinary purity," Sergio remarks, "marking the peak of design based on taut lines and plane surfaces. Very elegant, with

Pininfarina had planned to construct no more than five or six hundred Testarossas and ended up making more than seven thousand.

It was not necessary to see the Pininfarina signature to recognize the paternity of the Testarossa.

Ferrari 456 GT, 1992

Wanting to design a proper touring car, it was finally possible to bring the fiery horses back to the front: the Ferrari 456 GT was born.

exceptional performance, its streamlining is underscored by the distinctive design of the rear post that, without any interruption, joins up with the scallop that runs, waist high, the entire length of the side. It also gives easy access to the back seats, so that rear doors are not missed. In every respect a sedan, with the beauty and the lightness of line of a large coupé. We only produced around 4500 examples of the Fiat 130, it deserved a much better fate."

If Pininfarina from the very beginning has been a major protagonist in the success of the Italian line all over the world, what were the particular qualities of Italian style in those years of great triumph in all fields of industrial production? Mario Bellini writes in *Modo* no. 76, 1985: "Paradoxically, at the very peak of its popularity it may be said that there was no 'Italian style,' because Italian style was the absence of any definitive style; moreover, it had no unified code or stylistic elements that could easily be described or taught. And it is not by chance that this new 'Italian school' flourished on the lack of specific schools of industrial design, of furniture and of interior decoration." Again, in *Modo* no. 77, 1985: "This Italian style took its first steps in the 1950s with the Vespa, with Pininfarina's custom-built cars, the mythical Ferraris, with Nizzoli's Olivetti typewriters, with the optimistic transatlantic ocean liners, with the objects, the home furnishings and the exhibition preparations of Albini, Castiglioni, Gardella, Zanuso, etc. It was all a metaphor of a style without stylistic elements, made light and agile by the absence of a tradition that was recent enough to still be stifling, but still fed by the presence of a great tradition that was ancient enough to be transformed into a rich culture of behavior. This style first had to overcome the Scandinavian

The form of the Ferrari 456 GT is a contribution to the figurative culture of our time.

In 1993 a jury composed of distinguished personalities of the world of culture, at the Milan Art Pavilion, awarded the Ferrari 456 GT its prize for the most beautiful car in the world.

threat (which soon petered out in the impossible dream of donning the mantle of rationalist tradition in a craftsmanship-productionistic vein) to then assert itself in the mid-1960s as the style of no style, the style of research, invention, innovation; the style of typological, linguistic, technological experimentation and even as the style of anti-design, of radical design; the style of the return to styles, coming close at times to the worst of risks (though not as close as in many other countries)—namely, the involution of style-design. So this is how this Italian style grew and developed… with a taste and a culture of living and consuming that was more widespread, more refined, and more unabashed; with a fashion design and production that in the past decade asserted themselves as world leaders; with a gamut of publications on fashion, product design, furnishing and architecture that exceeds the sum total of what is published on the subject in the rest of the world… Italian style is not a style; it is a state of grace that has been holding up for more than thirty years; it is a leaven that has now spread throughout the lifestyle of the society that spawned it."

It is clear that this "Italian style," as Mario Bellini describes it, provides no codes of identification. Past and present, it has acted as a leaven in the batter of a full gamut of codes the world over. Therefore if, luckily, it is not possible to define the Ferrari Testarossa of 1984 with the words "Italian style," this car is unquestionably absolutely Italian, in the state of grace of Pininfarina. "The Testarossa," Sergio recalls, "was born when the BB had already asserted itself as daughter of the Berlinetta

456GT

previous pages
The extremely comfortable passenger compartment of the Ferrari 456 GT.

The large trapezoidal air intake that feeds the radiator and the engine compartment dominates the front.

Speciale of 1965, our first mid-engined research prototype. It was thus the fruit of all the experience matured in the mass-production of a model on a very small scale. We carried out long and meticulous studies in the wind tunnel to perfect the streamlining of the shape. We obtained very high down force values, ensuring maximum stability both in cornering and on the straightaway. The very name Testarossa evokes racing myth. It was a great success. We had planned to construct no more than five or six hundred cars and ended up making more than seven thousand. It represented the most advanced synthesis of its time, the felicitous reconciliation of the technical and the aesthetic. The design also called for the shifting of the engine-cooling radiator from the front to the middle of the car, according to the layout already tested in Formula One; this created a need for the large rear air intakes to which we added strakes which reinforced the car's powerful and aggressive image."

It was not necessary to see the Pininfarina signature to recognize the paternity of the Testarossa. The problem of the certificate of authenticity became urgent, however, in the broader field of products that fell under the general heading of "industrial design." Mario Bellini explains in *Domus* no. 726, 1991: "It is

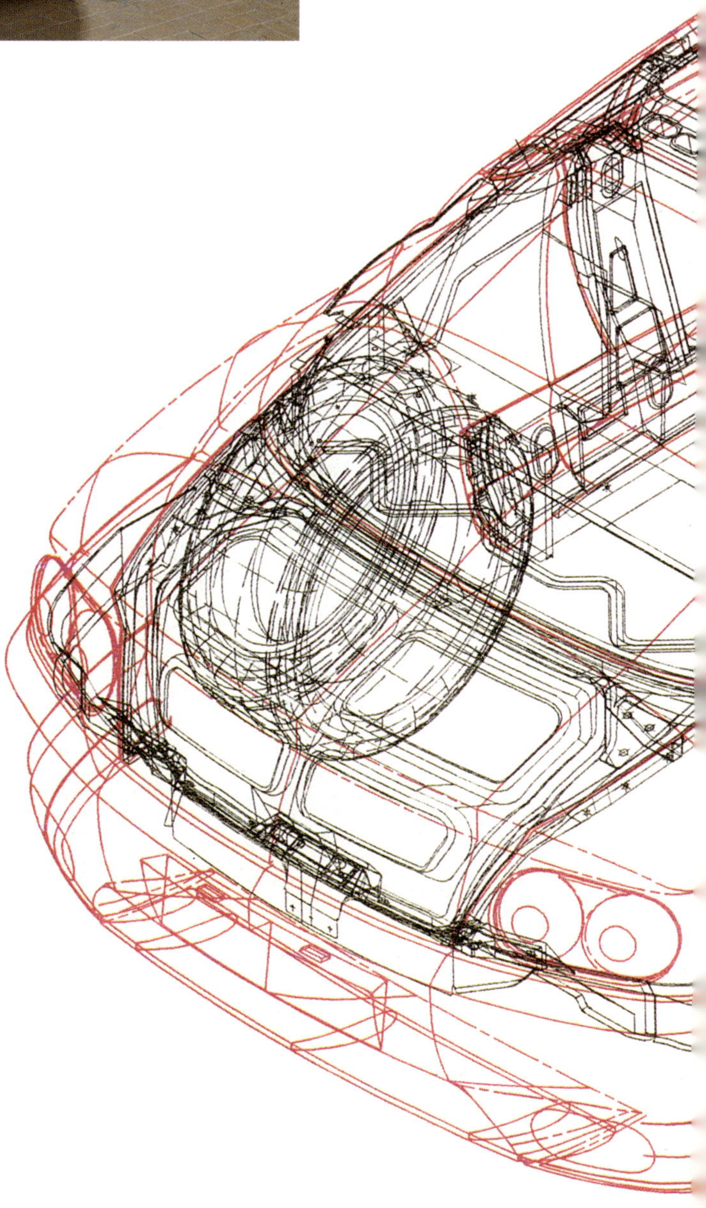

Ferrari 456 GT, 1992

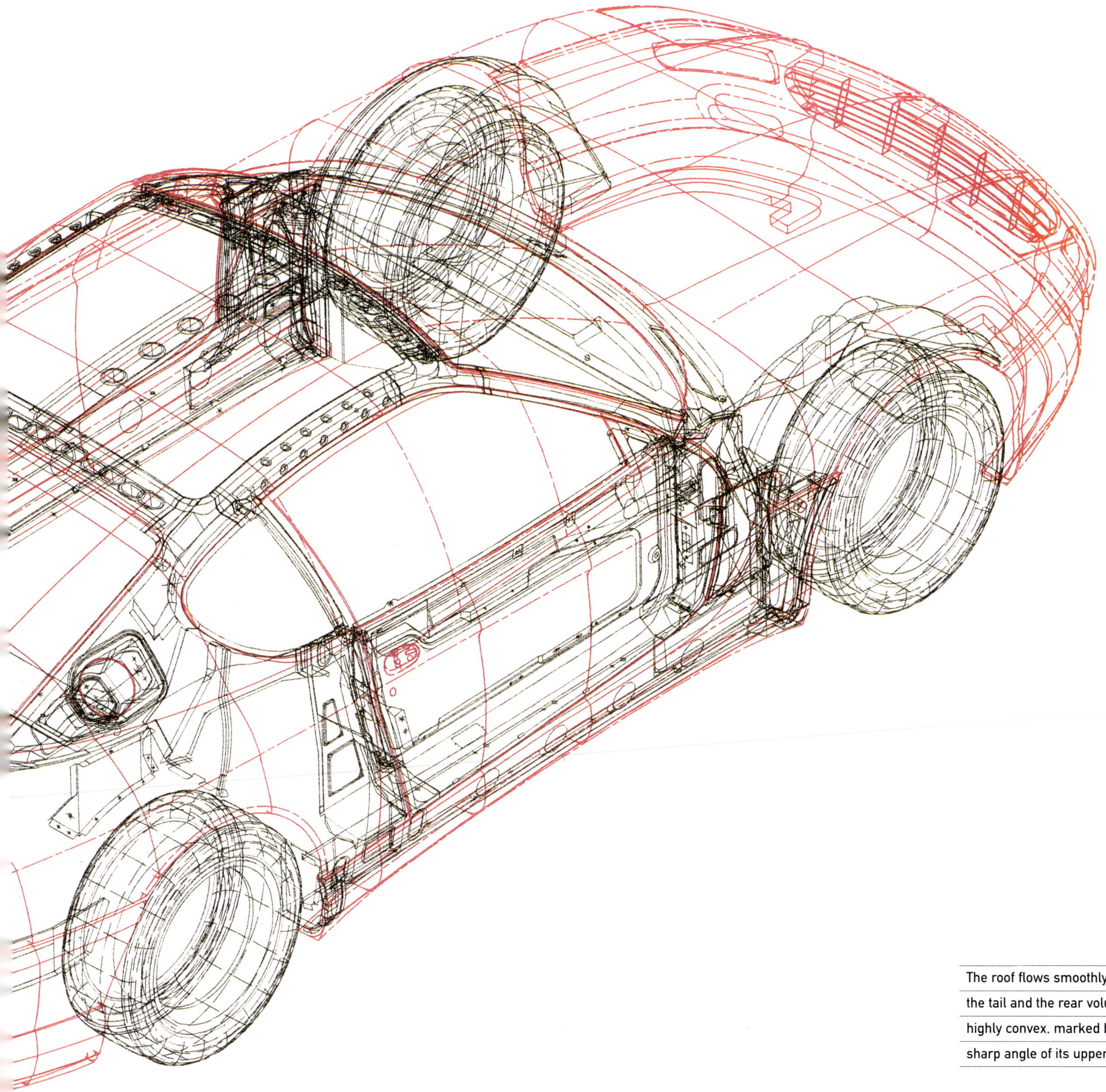

The roof flows smoothly into the tail and the rear volume is highly convex, marked by the sharp angle of its upper edge.

significant that this obsession with authenticity and with the signature should have exploded at the very time when it became clear that the heralded reorganization of the old workshops had been accomplished only halfway, imperfectly, and altogether unequally, while no genuine large industry had arisen to take their places after the crisis provoked over a hundred years earlier by the progressive and irreversible decline of manual skills. These obsessions, moreover, can be interpreted in two very different ways. On one hand they could express the persistence of an ingenuous faith in productive technological progress and in a new method of design inseparably linked to the charismatic and demiurgic figure of the industrial designer. On the other, precisely because of the untenabilty of this position in light of the substantial rejection of the industrial world by the culture of domestic space, we could interpret those obsessions as the regressive refuge in a post-craft and pre-industrial dimension in which the artist-author has taken the place of the craftsman-creator. It often happens that the logic of the signature leads to a forcing and distorting of the times and modes of creativity; it stimulates designers to worry more about their reputations than about their work; it feeds the circuit of Salons and of nonstop presentations of novelty, also to satisfy the asymmetrical thirst of the media and to boost the tactics of communication and sale. Very rarely is the author of a piece of furniture granted the luxury of a lengthy preliminary study, nor will he be easily allowed to disorient sellers and consumers with unexpected improvements and modifications after production has begun."

Sergio Pininfarina with the Peugeot 306 Cabriolet. Its structural rigidity makes the car elegant and sporty, but also quite comfortable.

San Giorgio, the finishing department: celebrating the first fifty thousand 306 Cabriolets produced.

A precise diagnosis: only the encounter of technology and aesthetics, the synthesis grasped in the instant of the evolutionary process, can truly be called a product of industrial design. The rest is form, and lays claim to the certificate of authenticity. If in 1968 the Daytona marked the limit of the reliability of a front-engined high performance car, and in 1984 the Testarossa exalted the efficiency of a mid-engined layout, in 1992 advanced traction-control technologies made it possible to opt for either one of the solutions with equal surety, according to the needs of the design. The mid-engined car, however, can seat two and no more. So wanting to design a proper touring car, a two plus two, it was finally possible to bring the fiery horses back to the front: the Ferrari 456 GT was born. "What struck me," Sergio recalls, "was the judgment of certain critics—real snap judgments: Pininfarina, after all, has simply redone the Daytona. If after twenty four years, at the birth of the 456, the Daytona still hadn't been forgotten, that was something for us to be proud of, not ashamed, all the more since we knew we'd outdone that car without forgetting its compactness, its pluck, the extreme elegance of its line." At the Art Pavilion of the Milan Triennale in 1993, a jury composed of distinguished personalities of the worlds of culture, art and architecture awarded the 456 GT the prestigious prize for "the most beautiful car in the world." The Jury judged its form to be a contribution to the figurative culture of our time. A similar statement had been made at the Museum of Modern Art, about the Cisitalia, some fifty years before. From the outside, what strikes one immediately are the curved surfaces marked by lines of tension and scalloped volumes that animate and lighten them. Even the sills and the bottom of the bumpers are body-colored, which makes the car look very close to the ground. The large trapezoidal air intake that feeds the radiator and the engine compartment dominates the front. The two lower intakes, for the brakes, are elements of the modeling of the spoiler area. Its size, more than the presence of designer elements, gives great importance to the hood, which integrates the retractable headlights. The front fenders accentuate the air vents of the engine compartment. The roof flows smoothly into the tail and the rear volume is highly convex, marked by the sharp angle of its upper edge; it houses two round tail lights. But that great sense of compactness, enchanting in the vision of the contours, does not prevent maximum passenger comfort.

If the synergetic relationship Pininfarina-Ferrari is at the heart of the glorious

Sergio Pininfarina with the Ferrari F50 at the 1995 Geneva Motor Show.

history of the Prancing Horse, what is the usual relationship between the large international car-makers and their designers? Sergio explains: "When a car company, and here we are talking about big companies, needs a new model and needs to renew the range of its products, a certain number of 'experts' prepares a briefing, which represents the spirit of the model or of the models to be built. If taken as a general indication, the briefing can be useful; if taken in an absolute sense, it makes the model downright unfeasible. Because it affirms a contradictory customer survey. It indicates the need to be all things to all people—fat, thin, tall, rich, poor, or whatever. So the 'experts' draw up this briefing for a product that can never be made. The feasible model *has* to disregard part of the survey, and whoever is running the show will be the author of these exclusions. To make the project feasible he *has* to take risks. So the designer consults the brief and extracts what he considers the most important points, then he makes his design, and then he puts up with the tedious criticisms of the 'experts.' We've had a relationship of design and production with our client Peugeot for fifty years now, and its still going strong. This is confirmed by the great number of models designed by us and produced by Peugeot, and the considerable number of niche models we designed and produced ourselves for Peugeot. The 306 Cabriolet, for example, which certainly does not belong to the biggest-selling category of the international automobile market. Still, we produced forty-two thousand cars between 1994 and 1997, and another ten thousand in 1998. In the five years since it appeared on the market, the 306 Cabriolet has reached a total of fifty-two

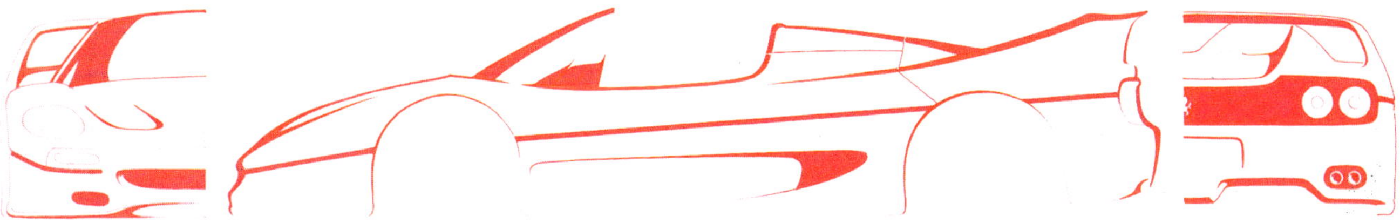

The F50 refers aesthetically to the great prototypes of the World Manufacturers' Championship: P2, P3, P4, up to the 352P.

following pages
The passenger compartment of the Ferrari F50.

thousand cars, and it's still in demand." For this car Peugeot assigned Pininfarina the study of the design, the engineering, and production. The Peugeot 306 Cabriolet was named in 1994, by an international jury, the best looking convertible of the year. Its structural rigidity makes the car elegant and sporty, but it's also quite comfortable. Ah! Those splendid French cars, open and blue, that traveled the dusty roads of the early years of the century! The open car has its new day, designed and produced by Pininfarina for Peugeot.

In 1995, for Ferrari's fiftieth anniversary, Pininfarina designed the F50, which sums up the grand sporting tradition of Ferrari in the style of their consolidated and fruitful collaboration. A 4698 cc, mid-engined V12 two-seater Berlinetta, the F50 is low, wide, and well

Ferrari F50, 1995

A great result, fruit of the consolidated synergetic relationship Pininfarina-Ferrari.

Sergio Pininfarina

Pininfarina designed and built the thoroughbred Maranello, opting for the front-mounted engine, and producing a far more powerful car than the Daytona and the Testarossa.

set on the ground with its huge low-profile tires. Aesthetically it refers to the great sports car prototypes of the World Manufacturers' Championship: P2, P3, P4, up to the 352P. It can take the form of a closed Berlinetta or an open Barchetta, thanks to two interchangeable bodywork parts. The car continued and perfected, in the wind tunnel, the thorough aerodynamic study of all the areas that make it devastatingly effective on the road. And then in 1996 came the Ferrari Maranello, ideal heir of the Testarossa of 1984. Thanks to traction-control technologies, engines may now be placed in the middle or in the front, with equally excellent results. In fact, the 456 GT was already on the road in 1992. Hence Pininfarina designed and built the thoroughbred Maranello with its front-mounted 485 horsepower, 5464 cc V12,

Ferrari 550 Maranello, 1996

Sergio Pininfarina with the Ferrari 550 Maranello, at the 1997 Geneva Motor Show.

producing a far more powerful car than the Daytona and the Testarossa. A great result, unquestionably fruit of the consolidated synergetic relationship Pininfarina-Ferrari, Sergio explains: "By ourselves we couldn't have worked miracles." A further example of a design that had no need of a certificate of authenticity: what is authentic is the relationship between Pininfarina and Ferrari. Imparting a sensation of tense power, the design of the 550 Maranello stretches the soft surfaces of the sheet metal with accentuated folds that create strength in all views of the car. Clear and precise volumes underline the excellence of the visual impact, exuding power and dynamism in the flanks traced by a dihedron that also marks the start of the rear fender. Between the wheelarch and the door, in the ample surface of the front fender, the placement the two hot-air vents of the engine compartment are a homage to the classical front-engined Berlinetta. The air intake for the engine's injection system is placed in the center of the large, softly modeled hood. The non-retractable headlight units and the tail-lights are integrated in the shape of the body. The interior provides both driver and passenger with the highest level of safety and ergonomics. As for the 456 GT, thirty years after the Daytona, the horses are pawing at the harness once again, and the driver can tranquilly and joyfully slacken the reins.

If for the Prancing Horse 1995 was the year of the 550 Maranello, for Peugeot it witnessed the birth of the 406 Coupé. "Coupés and cabriolets are part of the history of Peugeot and are at the root of its tradition," declares the official 1995 Peugeot catalog. The Peugeot coupés and cabriolets are thus part of the history of Pininfarina, with its characteristic respect for the corporate identity of its client's models and the consolidation of its traditional styling cues. But this collaboration with its historical partner "took on a new dimension with the 406 Coupé project. For the first time in fact Peugeot entrusted Pininfarina not only with responsibility for the exterior and interior form, but also with the industrialization and the construction, in strict respect of the norms, especially the quality norms, that Peugeot imposes daily in its own production." And the parameter "quality" spurred Pininfarina to take a close new look at the very Peugeot tradition that Pininfarina had done so much to consolidate. "There is a kinship with one of the coupés created for Peugeot by Pininfarina, the Coupé 504, which is revealed in the side panel creaseline, in the shape of the fenders that repeat it, and in the rear window that is straight

Vigorous and quick, the design of the Ferrari 550 Maranello stretches the soft surfaces of the sheet metal with accentuated folds that create strength in all views of the car.

1998 Car Design Award for the Peugeot 406 Coupé: in the photo we recognize Sergio and Andrea Pininfarina and Lorenzo Ramaciotti.

Exhibition at Sochaux in 1996.

and set back from a sharply angled post that underlines the car's third volume. This and other fine details are a bit like a *coup d'œil* that takes one back to the past of the Peugeot coupés, and contributes to an immediate recognition of the make." It is not only the Ferraris that have to be Ferrari in every epoch; the same criterion holds for the diverse identities of each and every client. This is the teaching that Pinin bequeathed to his son Sergio. "It thus becomes necessary to propose reassuring volumes, more serious and formal than ephemeral, but of high class and of an atemporal beauty that does not deteriorate rapidly in time." And the design is embodied. "In the plan view, the delicate balance of forms is created by a deliberate elongation of the hood, together with a slightly lesser rear elongation and the moderate height of the interior, to maintain the car's third volume, containing the trunk. On either side, the rubbing strip is deliberately shorter than the flank, to maintain a sense of continuity between the upper and the lower part of the sides of the car. This continuity is also to be found between all the surfaces of the body, whose soft and mobile forms are extremely sensitive to color. Each detail, such as the radiator grill and that which surrounds it, or the lower part of the body for example, contribute to the expression of a balance between the roundness and the tension of the forms, making for a car that is soft but of strong character. The importance Peugeot gave to this return to the coupé is underlined by the fact that, even though it originated from the 406 family, we conceived and wanted this new model to be a completely new car. In the study, in the construction of the prototypes, and in the phase of

Each detail of the 406 Coupé, such as the radiator grill and that which surrounds it, or the lower part of the body for example, contributes to the expression of a balance between the roundness and the tension of the forms, making for a car that is soft but of strong character.

industrialization, a great many problems posed by the complex form and demanding contents of the car have been faced and solved, at times with results that went beyond the original objectives. The strong and qualifying point of this model is at all events the body as a whole, which makes it possible to join excellent performance and roadability with extremely high safety levels. The level of the car's acoustic comfort is also remarkable. It was obtained through exacting tests carried out in the Pininfarina wind tunnel, aimed both at the reduction of external sources of noise and at the best possible design of a sealing system to minimize wind noise… The characteristics of the vehicle as a whole, in line with the most stringent parameters to be found in the auto industry, permit it to fulfill the highest expectations posed by a luxury niche model of great prestige like the Peugeot 406 Coupé."

In this car, then, Pininfarina fully satisfies those demands for the new that are necessary for the continuity of the Peugeot tradition. Powerful, sleek, streamlined, the 406 Coupé is, above all, *safe*. Ever since the 1960s safety has been Pininfarina's prime objective and the defining element of its ecology of car design. After all, the reconciliation of technology and

On either side, the rubbing strip is deliberately shorter than the flank, to maintain a sense of continuity between the upper and the lower part of the sides of the car.

aesthetics relates to the person who uses the machine and not the machine that uses the person. "In the first years of the century," Sergio explains, "the car was God and man was at the service of this Car God. That's not so today. Everything is at the service of man: I brake and the servo acts for me. There are even quiet Ferraris, with automatic transmission. I recall that Enzo Ferrari wanted his cars with light flywheels, and it was hard to shift gears. It wasn't just a whim; he wanted to exalt the lightness of the car. In the future, then, there will be more and more cars. The standard of living is destined to rise. We will all be richer and the super rich will be poorer. And at some point the car will also serve to put right all those negatives it has accumulated throughout its history. The problems of safety and of pollution will some day be resolved, but the car designer can only do so much. Society as a whole will have to take responsibility for the car, to deal with questions like city-planning, roads, traffic. But, more and more, it will be up to the designer to give form to technological progress by pursuing its technical-aesthetic synthesis. Niche products will preserve the memory of historical transitions; the car, in general, will be nothing more than the most advanced means of transportation. My father always believed that our little firm should act as a leaven in the big market of the automobile. This is even truer today, when it will be up to the niche product to bring back under man's control those very technologies that threaten to get out of hand. This, then, is the sense of our technical-aesthetic research at the service of the automobile, today like yesterday—the sense of our production: the continuous, exhaustive updating of the meaning of leavening."

In 1965 the Dino Berlinetta Speciale research prototype effectively began the history of Maranello's mid-engined thoroughbreds. Thirty-four years later, in 1999, the 360 Modena brings to light—unveils—the positioning of the engine as the distinguishing mark of the car in its totality. This mark is now the sign of a *choice*—not an indispensable design solution

Peugeot 406 Coupé, 1996

With the 406 Coupé, Peugeot entrusts Pininfarina not only with responsibility for the exterior and interior form, but also with the industrialization and the construction.

Peugeot 406 Coupé, 1996

The Ferrari 360 Modena, the "youngest daughter," so far, sums up the entire history of the Ferrari-Pininfarina synergetic understanding.

anymore, it unveils an aesthetic identity. "The choice of making the engine visible from the outside, through the rear window, was very important, enhancing the value of its aesthetic content and exalting its essential importance in a car of such extreme performance." So declares the text of the presentation of the Ferrari 360 Modena. "The new car is an important step forward in terms of both technology and form. It has two radiators positioned in front of the front wheels, so that the two air intakes are separated on either side of the front. From the functional standpoint this arrangement permits the further improvement of an aerodynamic efficiency that is already extremely refined in the treatment of the underbody. In fact the air flow at the front is not disturbed by the presence of a radiator in the center and the part that passes under the car further contributes to negative lift… This eminently technical and functional progress now presents itself in an innovative guise with respect to the classical form of the Ferrari with a central air intake; two separate air intakes had not been seen on a

Ferrari 360 Modena, 1999

Sergio Pininfarina at the Geneva Motor Show in February 1999 at the wheel of the 360 Modena.

The Ferrari 360 Modena is an important step forward in terms of both technology and form: it has two front radiators and two separated air intakes on either side of the front.

Ferrari since the 1.5-liter Formula One car and since the sports cars from the early 1960s. Today this set-up has been revived as a result of new and important aerodynamic progress." If traction-control technology makes the positioning of the engine in the middle possible, thereby transforming the design of the structure into an aesthetic element of the car, the most recent aerodynamic studies make further aesthetic solutions possible. "Other significant elements of the exterior design of this new Ferrari include the sharply pronounced fenders, the non-retractable lights as in the 550 Maranello, the air intake on the rear fender that ideally recalls the Ferrari 250 Le Mans, the symmetry of form between the front air intakes and the engine compartment air vent in the rear bumper… And then there is the use of aluminum in both the car's structure and in its interior, a lightweight element par excellence whose practical use goes back to the tradition of sports cars of the past, unlike those luxury cars in which wood was used as a purely decorative element." The 360 Modena—the "youngest daugther," so far—sums up the entire history of the Ferrari-Pininfarina synergetic understanding. A memory pregnant with meaning, that bears witness to the full evolutionary span of the two firms. "Ferrari and I, we both wanted to get the automobile away from the melodramatic, the monumental, even the visionary, and quit the decorative forever."

Ferrari 360 Modena, 1999

The comfortable interior of the Cadillac Jacqueline.

If even the dream has its name, it must be a really special car… Holder of only ten shares, Laura Partridge shakes up the top echelons of a big New York financial institution, getting herself elected by all the small shareholders to the board of directors. She makes some big changes indeed, and is rewarded for her efforts with the *Solid Gold Cadillac*, which gives its name to Richard Quine's film from 1956, with Judy Holliday in the role of Laura, Paul Douglas, Fred Clark, John Williams, Arthur O'Connell and Ray Collins. And a Cadillac enters the dream landscape of Bob Dylan's 1963 ballad *Talkin' World War III Blues*: "Well, I seen a Cadillac window uptown / And there was nobody aroun'. / I got into the driver's seat / And I drove down 42nd Street / In my Cadillac. / Good car to drive after a war. / Well, I remember seein' some ad, / So I turned on my Conelrad. / But I didn't pay my Con Ed bill, / So the radio didn't work so well. / Turned on my record player. / It was Rock-A-Day, Johnny singin', / 'Tell Your Ma, Tell Your Pa, / Our Loves Are Gonna Grow Ooh-wah, / Ooh-wah.'" The car was one of the symbols of the post-war American myth. Lorenzo Ramaciotti describes it this way: "In the late 1950s and early 1960s America *was* the car and life, with its drive-ins, revolved around it. It was a magic land where sizes were on a grand scale and industry applied methods of large scale production that everyone was trying to learn. Also the style that arrived from America had something mythical about it: the baroque redundancies of the forms, the hypertrophicity of the sizes, the absence of any research for the optimization of returns, weights, spaces. All this made stylists dream of measuring themselves against different schemes and conditionings—like the show cars

The unique model was called Jacqueline, in honor of the wife of the new President of the United States John Fitzgerald Kennedy.

A detail from the Cadillac Jacqueline: the twin headlights are quite high with respect to the hood.

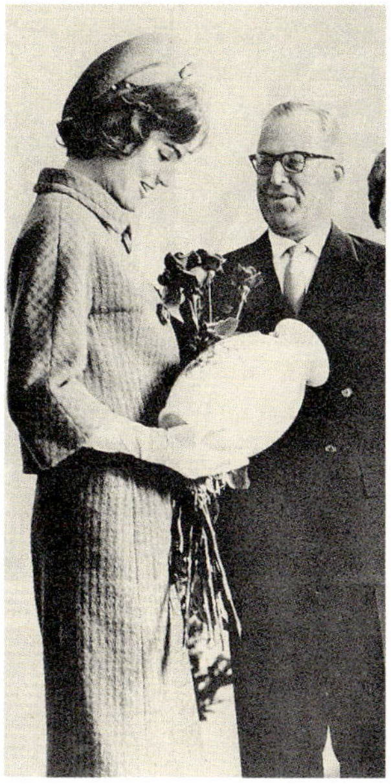

Jacqueline Kennedy in a photo of 1961. AFE, Archivio storico del cinema, Rome.

I seen a Cadillac window... I got into the driver's seat and I drove down 42nd Street.

Bob Dylan

of the Autoramas, as absurd and senseless as you like, but chockfull of wild ideas and unusual perspectives, like the excessive flavors of an exotic dish. For Pininfarina too the American myth was important, and General Motors a sought-after partner, especially when there were living legends like the charismatic and excessive stylist Bill Mitchell to be met with."

But the guide lines of Pininfarina design paid no homage to the stylistic status quo. This is why the 1961 Cadillac research prototype has to be considered a study on the possibility of redressing the gigantesque dimensions of the American car. The unique model was called Jacqueline, in honor of the wife of the new President of the United States John Fitzgerald Kennedy. And on the occasion of the First Lady's visit to Italy in 1962 the car was sent to the American Embassy in Rome. Pinin recalls in *Nato con l'automobile*: "The Ambassador invited me to a reception for just a few people. Mrs. Kennedy complimented us on our model, which she had examined in detail; she particularly liked the transparent roof. 'That way you don't miss any of the scenery,' she remarked. She spoke in French, out of politeness and to facilitate our conversation. She also said she would like to own and drive an Italian car. 'This seems to be an obsession of ours in America these days,' she added, 'but for me there's another reason: I love the classical quality of the shapes, the lines. But at the White House we only have American makes, and rightfully so.'" While the American myth was making inroads in Italy, our way of understanding the automobile fascinated the Americans even if, in the case of the First Lady, as she herself made clear, hers was a case of cultural empathy, of being fully in tune with the

The side of the Cadillac Jacqueline, marked by a simple scallop, gains sleekness from its proportions.

The Dino of 1965 was the first model Enzo Ferrari dedicated to his recently deceased son.

The Dino Berlinetta Speciale exhibited at the Le Mans Museum.

criterion of Italian design and, in particular, with that of Pininfarina. "The car had an aircraft-style transparent roof panel, lightened by the slenderness of the pillars and by the brushed steel surfaces. The side, marked by a simple scallop, gains sleekness from its proportions and is completed by a rear that is simple and rational." What does it mean to define a car as a research prototype? Sergio goes on to explain: "It is in the first place a unique model, generally but not necessarily fully working, where one concentrates the most advanced technical and aesthetic solutions of the moment so that, once tested, they may then be applied in regular mass production of the whole car or of certain parts. The Peugeot management, for example, wanted for their 204 a particular detail from the Cadillac Jacqueline: the twin headlights that were set quite high with respect to the hood."

We see, then, that apart from purely formal research one-offs, the one-of-a-kind prototype is used for feasibility studies for mass production, and the big international producers use it as a point of reference to pick up those details that best fit in with the identity of their particular make. But there are also prototypes that resolve specific needs, justified *a priori* by the urgency of a situation. This was the case with the Dino Berlinetta Speciale of 1965. The engine placed in the front in fact excessively lightens the back of the car at high speeds, casting doubt about its stability, and conditioning its clearance from the ground because of the exhaust pipes under the body. The research carried out by Pininfarina concentrated on the advantages offered by placing the engine in the middle of the car. "The Dino Berlinetta Speciale of 1965 also marked the changing of the guard at the

This car could be my granddaughter, it's a daughter of the second generation.

Pinin Farina

summit of our firm. The following year," Sergio recalls, "with the death of my father, I took over the presidency, and my brother-in-law Renzo Carli, with whom I had been working in perfect harmony for quite some time, became general manager. The prototype was presented at the Paris Salon and my father wanted to join me there at all costs, even though his health by then was very bad. He only stayed a few minutes, but long enough for a look at the flaming red Berlinetta and to pronounce the words that still make me proud today: 'This car could be my granddaughter, it's a daughter of the second generation.'" Pinin credits his son with having understood the identity of the design. As Lorenzo Ramaciotti notes, the Berlinetta Speciale of 1965 was "a revolutionary car which laid the foundations for the development of mid-engined Ferraris that have lasted over twenty years; for the production of the Dino 206, and then the 246 that derived from it, considered stylistically perfect. It was also a sharp break with the canons, renewing them without overturning them entirely." The Dino of 1965 was the first model Enzo Ferrari dedicated to his recently deceased son. "A very low hood, lenticular in shape, deeply recessed between sculpted, blended fenders that sloped down in front to four

In the research prototype Dino Berlinetta Speciale of 1965, study and invention will make possible the mass production of an entire family of cars.

The Dino Berlinetta Speciale is a revolutionary car, which lays the foundations for the development of the mid-engined Ferrari.

headlamps, behind a plexiglas fairing that continued the bodylines…" The description is highly suggestive. "The body side carries to the extreme the theme of the dihedral that here, thanks to the distinctively prominent volumes, becomes a sharp rift that nearly divides the car into two halves. An element appears that was to become a typical and recurrent theme: the scalloped air intake on the door and the rear fender, which is here traversed by a chrome bar acting as a handle… To balance the importance of the space of the front fenders, which arise naturally from the surfaces of the body side and the hood, the flank is slightly marked by the motion of the rear fender, which interrupts the waistline… To solve the difficult equation between function (which requires the engine compartment outside the passenger compartment for reasons of acoustic and thermal insulation, a vertical rear window for visibility, and easy access to the mechanical organs) and aesthetics (which require a smooth transition between roof and tail), the amphitheater rear window was invented, which joins up with the fin giving lightness and transparency to the rear pillars and satisfying all the technical requirements. This solution has become a constant of all the rear-engined Ferraris designed by Pininfarina up to the Testarossa."

This description of Lorenzo Ramaciotti's gives us a good idea of what is meant by a research prototype, in which study and invention make possible the mass production of an entire family of cars.
The synergetic relationship Pininfarina-Ferrari progressively reinforces the distinctive image of the cars of Maranello through a series of models

In the rear of the Dino Berlinetta Speciale we see the amphitheatrical rear window, which joins up with the fin giving lightness and transparency to the posts. Note the reference to the 375 Mille Miglia of 1954, where the form of the rear window is a matter of pure aesthetics.

that make up its tradition, each model an episode of the common intention of their design research. Hence these models represent objectives attained and sealed by this felicitous correspondence of design. In this regard the Testarossa of 1984 and the 550 Maranello of 1996 are emblematic. If the two cars are diametrically opposite in conception, that is certainly not the case where the coherence of research of the two design partners is concerned, with the result that both cars can be recognized as Ferraris and only Ferraris. Thus in time, and on roads throughout the world, Ferraris make Ferrari great and make Pininfarina great.

It has been an entirely different story, however, with the niche rival, Porsche. Just one single model, the 911, assures the Stuttgart car-maker its place in history. Porsche, then, corresponds to a single line of research aimed at the perfection of the 911 project. I wrote in my article "La Porsche evoluzione design" in the column "Nautilus" in *Abitare* no. 379, December 1998: "Porsche intends to affirm, in the 911, a criterion of design that aims to fulfill the expectations of the many. Even if, in this specific case, 'the many' consist of the particular niche of speed worshippers who, in their turn, are part of 'the many'—they are certainly not professional racing drivers. Where racing drivers are concerned, Porsche has been fulfilling their expectations ever since the days of the mid-engined 16-cylinder Auto Union with which Tazio Nuvolari, the 'Flying Mantuan' as D'Annunzio called him, won a number of races. Then, with Volkswagen, the 'people's car,' it even went beyond the concept of niche, offering the many a medium-performance means of transportation. But it must not be forgotten that

Dino Competizione, 1967

Wooden mock-up of the Dino Competizione.

In 1967 Pininfarina built a prototype with a decidedly race-course flair: the Dino Competizione.

the 911 bears the Porsche name and must therefore be different in every respect, must affirm its law." Porsche took up a challenge "and the word 'challenge' allows us to better understand that declaration of identity of the 911, made as only Porsche could have made it, when the car first came out in 1963. 'Having opted for a very personal conception of the car, characterized by a horizontal air-cooled 6-cylinder engine, positioned overhanging the rear axle, we certainly did not choose the easiest solution.'" Certainly the Dino Berlinetta Speciale of 1965 made no claim, as the 911 did, to be a difficult theorem to prove; it was simply the archetype of the evolutionary necessity of the Ferrari line at that moment. "Porsche wanted the immediate complicity of the user; then, in time, it did its best to lighten the weight of the user's irrevocable choice. The myth of the 911 stems, then, from the encounter with the user who becomes enamored of its evolution. And even if, over the years, each dated model is perfect nonetheless, Porsche shall never cease to pursue the dialectics of quality... so that the first model, like the last model chronologically, in the Porsche criterion can never be the last or the first." It is clear here that technical and aesthetic perfection is only at the service of the proof of the theorem 911. "The body must not undergo substantial changes, it will always take part in the improvement of the car in its totality. Thus in 1971 moving the rear wheelarch back underlies the constructor's decision to obtain the greatest possible balance of the car's driving position; but the modification takes on the form of its function, and the body benefits in compactness and streamlining. Since 1968 the transition from the use of steel wheels to alloy wheels and to more and more sophisticated alloys, while

The doors open like seagull wings, permitting easy access to the interior.

lending the body greater visual impact, has also made it possible to improve brake cooling. And fitting wider tires made the car more stable and was permitted by the wider fenders. Until 1989, the greatest modifications concerned the adoption of higher and larger bumpers, required by the norms of the American market, which involved reinforcements in the interior of the body. The latest model of the 911, from 1996, as always in the Porsche criterion, is ancient with a modern flavor. Sure, the greenhouse is larger than ever before, there's plenty of cabin space, the line is clearly streamlined and also compact, harmonious. But it is still and above all the 911: it is not the first, nor will it be the last."

In the same years that marked the evolution of the 911, the history of Ferrari was a chorus of voices—all of them 'primadonnas'—called the 250 SWB of 1960, the 365 GTB 4 Daytona of 1968, the Testarossa of 1984, the 550 Maranello of 1996, to mention a few. The users of that privileged niche which make all car-lovers dream will declare themselves, on one side, enthusiastic supporters of the evolutionary theorem Porsche 911, on the other, of the evolutionary dialectic of the Ferrari models. "With the Dino prototype of 1965," Lorenzo Ramaciotti notes, "Pininfarina clearly stated the aesthetic canons by which it was to develop the mid-engined Ferrari cars." It had thus made explicit the theme of development of a particular category of cars. Now was the time to get on with the task. Thus in 1967 Pininfarina built a prototype with a decidedly race-course flair, which it called Dino Competizione [Competition]. "On the car two cantilevered wings have been installed, one in front, the other in the rear. The two appendages can be adjusted to optimize the balance of the car, and they are an absolute first on a sports prototype with covered wheels… The aesthetic research smoothly asserts itself in the abolition of the front air intake, in the bare hint of a fin that livens up the rear of the car, and in the sharp creaseline that livens up the body side." Sergio recalls: "Ferrari was diffident at first… He considered me a young fellow long on enthusiasm but very short on experience, and he wasn't sure about my competence. Besides which, he was used to dealing—and always in the first person—with the highest levels. But he was also a nonconformist, a person who thought for himself. And maybe he was amused by the idea of dealing with this young fellow with the candor and imagination of a boy who was not yet polished by the vicissitudes of life and business. Anyhow, almost secretively, covered by that veil

The aesthetic research smoothly asserts itself in the abolition of the front air intake and in the bare hint of a fin that livens up the rear of the car.

The passenger compartment of the Alfa Romeo 33, with the doors open like seagull wings.

Accepting the proposal of the Alfa Romeo president Luraghi, Pininfarina in 1969 produced the Alfa Romeo 33, with real racing car flair!

The line of the front of the Alfa 33 is concave and joins together the prominent volumes of the fenders.

of reserve and modesty typical of those men like Ferrari and like my father who were born in the past century, little by little he gave me his esteem and then his affection. At my father's funeral, unexpectedly Ferrari said to me, 'From now on let's address each other as *tu*.' That was something that meant very very much; it expresses especially his full confidence in our work together and in my future."

If physiognomics or physiognomy deduces the peculiarities of a person's character from their bodily appearance, studying coincidences with and distances from an average model represented by presumed constants, physiognomics applied to the automobile deduces from the techno-aesthetic appearance of a unique research prototype the coincidences with or distances from mass-produced models, especially those that are produced in great numbers. Indeed, the prototype is of fundamental importance in the declaration of identity of a design's intentions. "It can investigate, learn to know," Pierre-Yves Thomas notes, "as far as it can possibly go, exceeding the limits of the normal, of the acceptable... In the late 1960s and early 1970s the distance from the norm was at its peak; never before in automotive history had the formal invention of prototypes moved so far away from the aesthetics of mass production... Furthermore, if we make reference to the chronology of Pininfarina, we realize that the period of freest perspective corresponds to the period of the

Alfa Romeo 33 Coupé, 1969

Sleek line, low and very thin front, high tail: these are some of the characteristics of the Ferrari 512 S.

firm's technological development (1966–1972)." Pininfarina designers had at their disposal in that period, for the first time, their own state-of-the-art testing instruments. The use Pininfarina made of these instruments was decisive for its way of understanding the meaning of its research. Its prototypes, even the unfeasible ones, were to exalt this meaning to the advantage of the most judicious mass production, not in coincidence with the techno-aesthetic constants of the big car-makers.

As Sergio recalls, "Luraghi, the very sporty president of Alfa Romeo, said to me one day, 'Why only Ferraris, why don't you try with Alfa too? You have to make us an Alfa with real racing car flair!' The mechanicals of the 33 lent itself to it, so we accepted the challenge." The Alfa Romeo 33 Coupé research prototype is from 1969. Lorenzo Ramaciotti notes: "The forms are extremely plastic, their inspiration is nearly organic and may be attributed to the aerodynamic matrix of many Pininfarina studies,

The entire roof panel of the Ferrari 512 S tilts forward for access to the interior.

The side of the 512 S is decidedly flat, almost graphic, distinctly wedge-shaped in the lines of the front fender.

but also to an aesthetic choice that privileged more sculptured forms, an all-rounded look, rather than origamis obtained with adjacent bent surfaces. The concave line of the front that joins together the prominent volumes of the fenders gives, at the same time, aerodynamic efficiency and aesthetic novelty. The passage between fenders and flanks are sections in continuous and fluid evolution, without monotony and without abrupt breaks… The compactness of the cockpit, a crystal teardrop with light pillarwork, reinforces the feeling of power emitted by the body of the car." Alfa Romeo occupied, from the start, a special place in the design interests of Pininfarina, which was born thanks to the stimulus and the contribution of Vincenzo Lancia. "It's clear that for my father Lancia was like a wife, while our relationship with Alfa Romeo was like with a lover. I remember in fact when we went together to Milan to Alfa, we almost felt a little guilty." (from Luca Ciferri's interview with Sergio Pininfarina).

There were no guilt feelings however where Ferrari, peerless, was concerned; the rivalry was between Lancia and Alfa, "lancisti" and "alfisti." The 4193 cc, rear-engined, V12 Ferrari 512 S research prototype dates from 1969. Lorenzo Ramaciotti notes: "The car seems to be composed of a two-dimensional side panel that interprets the distribution of the masses, and an upper part on which all the modeling is developed, with the two sections joined by a sharp edge marking their perimeter. The side, with a creaseline low down, is decidedly flat, almost graphic, distinctly wedge-shaped in the lines of the front fender, the waistline, and the lower part of the rear fender… This dynamic and soft line simply and naturally links the low

Is it still an automobile, or is it the service module of a spaceship of 2001: A Space Odyssey?

Lorenzo Ramaciotti

and very thin, almost nonexistent front end with the high tail, indicating the position of the engine by enlarging the masses. This perimeter line, which is quite ample due to the vertical development of the side, acts as the base of a smooth line which rises without interruption from the hood to the windshield, set off by the fender bulges and the deep tunneling of the air intakes: the windshield and the slits of the air vents over the engine are an elegant graphic decoration that underline the softness of the surface. The rift with the everyday is emphasized by the way the entire top of the car opens up like a shell for access to the interior."

In 1967 Italo Calvino (1923-1985) published *Ti con zero* [t zero]. In one story, "L'inseguimento" [Pursuit], the author portrays a man unable to escape from the condition of the pursuer pursued by existential restraints. But here, the theater of action or of the impossibility of action is city traffic. Beyond the metaphor of Calvino's story we also find those problems which can no longer be solved in the pleasure of your car's performance. Problems that in the future will concern, and more so all the time, formal techno-ecological contrivances, not to mention the city-planning resolutions to apply them. "The car pursuing me is faster than mine… In my flight I head for the center of town; it was a timely decision; the pursuer is still behind me but now we are separated by several other cars; we're stopped at a traffic light in a long line…" It is sufficient that both pursued and pursuer bury themselves in traffic for the advantage—in terms of performance—of the latter's car to be annulled. The most impetuous thoroughbreds are penalized just like the slowest cart-horses. "The problem in sum is whether the line is divisible into a series of segments each with a life of its own or whether it has to be considered a single and indissoluble body, in which the only change that can be

The Modulo of 1970 is far closer to the evocative, beautiful object than it is to the most extreme example of a means of transportation.

Sergio Pininfarina, Renzo Carli and Gio Ponti with the Modulo of 1970.

expected is a decreasing of the density at nighttime, until a limit of rarefaction is reached in which only our two cars will conserve the same direction and will tend to annul the distance… What our calculations undoubtedly have in common is that, in both, the elements that determine the individual motion of our cars—power of the engines and ability of the drivers—count hardly at all, and that which decides everything is the general motion of the line, or rather the combined motion of the various lines that intersect in the town." The pursued will find no way out of this labyrinth of lined-up cars, he will remain "a prisoner of the general system of moving cars, in which no distinction is made between pursuers and pursued."

This is an interesting observation when, precisely in the years of "car wars," Pininfarina research prototypes took their distance, as never before, from that mass production standard we call "traffic." Thus the Ferrari Modulo of 1970 is far closer to the evocative, beautiful object than it is to the most extreme example of a means of transportation. The Modulo utilizes the same chassis as the 512 S. "A single volume, it stems from the superimposition of two symmetrical shell-halves," Lorenzo Ramaciotti notes, "joined in the middle by a perimetrical cut. The volume of the shell is ribbed by two longitudinal dihedrals that mark the passage from the hood-windshield-roof to the side, while the midsection is a taut and continuous, but greatly flattened, arch. The volume is sharply truncated at the rear by a perfectly vertical surface. Two cylindrical surfaces, housing the rear tires, are grafted onto this simple, immediately evident volume. Here we have no subtle attempts at sculpturing. The power of the object is in its geometry, the rest is decor—the color, the large hexagonal side windows that unite the glazing and air intake, the empty compartments above the front and rear wheels." And how does the Modulo relate to those mass-produced models that invade byways and highways? "Is it still an automobile, or is it the service module of a spaceship of *2001: A Space Odyssey* that takes off thanks to anti-gravitational lift and not on four wheels moved by an internal-combustion engine?" A curious observation, if one recalls that the automobile, half a century before the Modulo, still a child, was in pursuit—on earth and in the dust of the earth—of the myth of speed. That myth had been exalted by poets of the pen and of the wheel, the Flying Mantuan among them—Tazio Nuvolari, who had just

A phase of the beating of sheet metal on a template; a step in the assembly of the body, at the research center in Grugliasco.

turned twenty-four. "The race-course is truly in the driver's mind—the layout, the geometrical construction, even before he takes the road... The test concerns the directions, especially the curves. He *feels* them—and *how* he feels them, in his innermost depths, right in his bones... How else can you be sure you're tracing the right line? And the tires bite the apex, and you feel glued to your seat, and you come right out into the straightaway, ready to attack the next curve." (My portrait of Nuvolari, in *Capital*, February 1997).

Filippo Tommaso Marinetti, in the first issue of *L'Italia futurista,* wrote an article with the title "The New Moral Religion of Speed" (Milan, 11 May 1916): "Man began by denigrating the isochronous and cadenced rhythm of the great rivers that was identical to the rhythm of his own step. Man envied the rhythm of streams, similar to the rhythm of a galloping horse. Man tamed horses, elephants and camels in order to manifest his divine authority through an increase of speed. He formed alliances with the most docile animals, captured the rebellious animals, and fed upon the edible animals. Man stole electricity from space and fuels to make new allies for himself in engines. Man forced metals conquered and made flexible through fire to form alliances with fuels and electricity. Thus he formed an army of slaves, hostile and dangerous but sufficiently tamed that they transport him swiftly on the curves of the earth... Man tames material, metals, and he extracts machinery and forms machinery: trains and automobiles... Twisting paths, roads that follow the sluggishness of rivers and wind along the uneven backs and bellies of mountains—these are the laws of the earth. Never a straight line; always arabesques and zig-zags. Speed at last gives human life one of the

A striking photo of the Modulo in Mexico City, July 1971.

The XJS Spider of 1978 can boast of Pininfarina's experience and close study of aerodynamics, five years after the inauguration of the wind tunnel.

characters of dignity: the straight line.
The Danube, opaque under its tunic of mud, its face bent over its inner life full of libidinous and prolific fat fish, passes muttering among the implacable high cliffs of its mountains, as in the immense central corridor of the earth, a cloister unroofed by the swift wheels of the constellations. For how long will this donnish river permit a car to overtake it at top speed, with its barking of foolish fox terriers? I hope to see the Danube racing soon in a straight line at three hundred kilometers an hour." Luckily Marinetti's delirious invocation will not disturb the Danube's course and, a half-century later, in 1968, Stanley Kubrick will accompany the most evocative scenes of his *2001: A Space Odyssey* with the music of a great son of the slow and muttering river, Johann Strauss. Perhaps the Modulo object, so evocative but so pitiless in its detachment from the reality of the automobile, would have cut a fine figure to the sound of that music in Kubrick's space dance.

The Modulo represents, then, the peak distance from the norm on the curve, calculated by Pierre-Yves Thomas, of the sequence of Pininfarina research prototypes in the period from 1952 to 1990, although there was a sharp return toward the norm beginning in the mid-1970s. The 5343 cc, 12-cylinder Jaguar XJS Spider, dates from 1978. The prototype aims both at continuing and renewing the distinguished image of the prestigious British car-maker. Piero Casucci notes in *Jaguar, il fascino di un nome* [Jaguar, the Fascination of a Name]: "The car that, more than any other, has aroused the curiosity and interest of the English, and especially of the press, is the XJS 'spider.' The reason for this success is strictly related to

From an aerodynamic standpoint the XJS Spider is very clean, particularly beneath the floorboard.

Bob Blake

The XJS Spider is the car that, more than any other, has aroused the curiosity and interest of the English. It was in fact Pininfarina's intention to draw on the great traditions of the D- and E-Type Jaguars.

the goals Pininfarina set out to attain: to draw on the great traditions of the D- and E-Type Jaguars by opting for the rounded lines that had characterized their image while maintaining the unmistakable oval shape of the front air intake as a formal reference to the Jaguar tradition… 'I like some aspects very much,' Bob Blake affirms, 'especially since we're dealing with an open car. From an aerodynamic standpoint—a subject I've learned a lot about thanks to Malcolm Sayer and the D-Type—the XJS Spider is very clean, particularly beneath the floorboard. And no use is made of a spoiler or of wings. This is the secret of how to eliminate aerodymanic resistance by letting air flow under the car.'" Five years had gone by since the inauguration of the wind tunnel and the XJS Spider can boast of its experience and close study. "Today it would be called an organic design," Lorenzo Ramaciotti notes. "The tail, of a normal size for today, at the time was most audaciously large, and even the windshield with its concealed pillars was a novelty. The doubled top proposes a compromise between the safety of the roll bar and the freedom of the open car."

Twenty-four years after the presentation of the one-off Dino Berlinetta Speciale research prototype of 1965, Pininfarina built the Ferrari Mythos in 1989. Five years have passed since the debut of the Testarossa series and the Mythos takes over its tried and tested running gear. "A car of great technical and formal results," Sergio declares. "The Mythos is composed of two cars one inside the other. The rear, stronger, larger, carries the engine; the front, streamlined, cuts through, penetrates." Augusto Morello examines the topological implications of its formal treatment: "With the two rectangular tori in the plan, inserted one inside the other, it recalls the disquieting topology of Klein's bottle. It's a text-book example of how to define a Gestalt—namely Form with a capital F; that is, not just 'shape' or mere aesthetic stimulus, but Form, connected with the world and not only with the means. In that sense, Mythos, a Ferrari, is the most recent epiphany of an adventure that began in a *trattoria* in Tortona, an interpretation that would have seemed a very new automobile phenotype even to Pinin and Ferrari, champions of realistic utopism." The Mythos, even with its great technical and aesthetic innovation, still

With the two rectangular tori in the plan, inserted one inside the other, it recalls the disquieting topology of Klein's bottle.

Augusto Morello

remains a car, and for this very reason it is a Ferrari as Pinin Farina and the Drake would have wanted it. But the Mythos goes far beyond the Testarossa, and in this sense is even more Ferrari than the most famous Ferrari of the day. Ramaciotti in fact notes: "The Ferrari Testarossa mechanicals are an ideal starting point for formal research of this type. The exceptionally high performance, the mid-engined architecture with side-mounted radiators, the charisma of the Prancing Horse, are all a chance and a stimulus to *dare*… In the development of the project the technical and functional contents have not been sacrificed to the search for a strong formal presence: two mobile aerodynamic addenda, a front lip and a rear wing, move into working position when a predetermined speed is reached. The removable body is constructed in composites, according to the construction techniques used in racing cars. The car in motion, on the road, is the demonstration of an equilibrium between aesthetic emotion and technical rationality."

If in 1961 the Cadillac Jacqueline research prototype, compared to the gigantesque American cars, suggested balance, harmony and dash, the Chronos, thirty years later, in 1991, took up the demand to match high quality performance with equally high quality aesthetics. Especially when, even more than the sedan, its symbol will be the coupé. Pininfarina designed and constructed a two-seater on General Motors' 3615 cc, in-line 6-cylinder Opel Lotus Omega. Decidedly sporty, the prototype fully accomplishes the intention of the design. The official presentation explains: "From the standpoint of form, the model presents itself as a sculptural body plastically composed by the

Detail of the air intake of the Mythos prototype of 1989.

The Mythos is composed of two cars one inside the other. The rear, stronger, larger, carries the engine; the front, streamlined, cuts through, penetrates.

Two architectures are confronted: the Mythos and the Palazzo Te in Mantua.

The rear wing moves into working position when a predetermined speed is reached.

The body of the Ferrari Mythos is constructed in composites and is completely removable, according to the construction techniques used in racing cars.

integration of two aesthetically similar volumes: the first, low and broad, covers the wheels, and the second encloses the mechanicals and the interior… The width of the body side is emphasized by the reduced transverse dimensions of the top, whose tapering form is in harmony with the tapered design of the car as a whole and increases its muscularity. As for the side, a sharp stroke of the spatula traversing the entire side panel helps to lighten up the volume. Below this line, the side is deliberately sober, animated by a slight dihedron that runs around the car's entire perimeter and, widening toward the bottom to fair in the wheels and the side members, comes to create an optical support for its mass… As a whole the classical convex form of the flank's creaseline is enlivened by a succession of full and concave lines."

The Chronos confirms, splendidly, Pininfarina's experience in the world of the coupé, at the service of the best running gear from all over the world, including that of General Motors' Opel Lotus Omega. Indeed, as regards niche products—coupés, cabriolets, "spiders"—Pininfarina left a series of indelible marks on the history of the automobile. And so a small, compact open car, the Barchetta, was to be the first prototype of the 1992–1995 Ethos research project. "Ethos" declares the intention of its design: to be as mindful as is realistically possible of an ecological definition of the automotive product. Augusto Morello writes: "Ethos expresses the awareness of the fact that, along with the evocative wealth of myth and the metaphors attached to the product, there are now environmental restraints that take on cultural value only in the overcoming of the product itself, in the capacity to make it become not an enemy but an active part of a livable world. With no less intensity than Mythos, but replacing metaphor with allegory, Ethos (already followed by other works in progress) proposes a different syntactics

Ferrari Mythos, 1989

The width of the body side is emphasized by the reduced transverse dimensions of the top, whose tapering form is in harmony with the tapered design of the car as a whole.

The Ethos 3: preparation of the wooden mock-up and wind tunnel test.

relativized, and the multiple driver solution further demythologizes the tradition meaning of "ownership" of the car.

Ethos, Ethos 2, and Ethos 3 promote the functional aspect of the automobile, overcoming the most abused of its connotations and hypothesizing new ones. Thus when Pininfarina constructed the Honda Argento Vivo [Quicksilver] in 1995, the triad of the Ethos Project had already yielded up its fruits, which went to the advantage of further evolution. "The forms of Argento Vivo," its presentation declares, "enclose a structure in extruded aluminum integrated by specific joints, also cast in aluminum." The structure is therefore light, rigid, resistant, recyclable. "The architecture of the car is classical, with the 2500 cc 5-cylinder engine placed in the front, longitudinally... Aluminum, which with the basic structure constitutes the backbone and the heart of the new Pininfarina 'spider,' plays a fundamental role in the play of the two-tone finish and in the enlivening of the surfaces. The hood and the trunk in polished aluminum, the structure of the roof, the air intakes, are all constructed with this material that calls to mind sports or racing cars of great class and, at the same time, offers a sense of solidity, of protection, of functional refinement. The rest of the body is in blue-painted resin, a sort of support in which the parts in aluminum are set." Aluminum, then, extends its function of supporting structure and participates in the configuration of the car together with the resin that compacts the body. "In the closed version, the Argento Vivo is striking particularly for its glass roof, which together with the rear window forms a single transparent surface... Jointly operated by small electric motors and hydraulic stretchers, the roof

Above, the Ethos prototype; below, the Ethos 2. With this Project, Pininfarina is thinking of a sports car in terms of fast acceleration and not of high speed, whose components are completely recyclable.

Sergio Pininfarina shows Gianni Agnelli the Ethos 3 at the 1994 Turin Motor Show.

first is placed against the rear window and then pivots, together with its structure, into a compartment situated between the cockpit and the trunk. An electronically operated cover then encloses the compartment, creating a continuity of surfaces as in the Barchetta." The high technical and aesthetic value of the car is clearly revealed in the movement of this fully retractable glass roof, which in its splendid dynamics becomes the form of its function. Analogously, wood is both form and structure of the interior. Aluminum, resin, wood, glass: noble materials working together to give form, color and content to this "spider," whose very special colors are exalted even in the ring of its name: Argento Vivo [Quicksilver].

Forty years earlier a car that has since become a legend made its debut, with a sweet name of silvery timbre: Alfa Romeo Giulietta Spider. Argento Vivo and Giulietta, today placed side by side, very well explain the technological leap that separates them; but, no less well, Argento Vivo is able to explain its function as a prototype, the belonging of both cars to the same intention of design, to that particular way of understanding the mobile object which has always distinguished Pininfarina in the quality of its techno-formal solutions. "Sergio often speaks of beauty," Augusto Morello writes, "but when you go deeply into the analysis of his creations and of others', you realize how what he attributes to the automobile, when he does so, is actually part of a judgment referred to the quality with which the complexity has been resolved, an expressive synthesis that takes into account not only the form of things but the form of the world... Design is for him—and he preaches this—only the last phase of a deeply lived

Ethos 2 at the exhibition in Seoul in 1998: "Civilization, city and car. Pininfarina: from Leonard to the future."

following pages
One of the most spectacular solutions of the Ethos 3 is the possibility of driving the car from three different positions: righthand-drive, lefthand-drive, center drive.

Lorenza and Andrea Pininfarina with the Ethos 3; in the background, the Ethos 2.

Peter Ustinov, Bernard Cahier and Sergio Pininfarina examine the Honda Argento Vivo at the 1997 Geneva Motor Show.

intelligence." And it is precisely this "lived intelligence" that becomes a way of understanding form.

Thus for the CNR Eta Beta of 1996, as the official presentation explains: "The conceptual approach has preceded the aesthetic one, which was only dealt with later on, and with the sole purpose of making the result more pleasant and soft, fun and different, by playing on the destructuration of the surfaces underlined by side panels with parallel lines tilting toward the front, devoid of linear continuity and of any reference to the classical horizontal waistline." What is, then, Eta Beta? "The attempt to break with the concept of the automobile seen as a monolith had already begun with the Honda Argento Vivo project in 1995, where the diversity of construction materials—polished aluminum, painted resin and supporting wood—participated in the disassembling of the form, enhancing the value of the design. With CNR Eta Beta the tendency is heightened: it is by optimizing the difference of form and of the surface treatment of the components—the aluminum of the doors, of the roof, of the structures, the thermoplastic painted resins—

The glass roof together with the rear window forms a single transparent surface.

Honda Argento Vivo, 1995

The hood and the trunk of the Honda Argento Vivo are in polished aluminum; the rest of the body is in blue-painted resin.

following pages
Sketching the birth of the Argento Vivo.

that the car is interpreted." The components of the car are thus understood as structure, form and color working together for the final result. The prototype can be interpreted in the individual techno-aesthetic solutions that make up the MPV, and it is these solutions that make up the definition of research prototype. "Eta Beta is fitted with a hybrid engine system (a front-mounted, 1800 cc 4-cylinder 4-stroke internal combustion engine, and coupled electric motors with permanent magnet and axial flow in the rear wheels) and presents itself as ecological, totally recyclable, safe, ergonomic, with a modular interior and is thus suitable both for city and for country driving." The car, that is, can make use of its electric (zero emission) motors in the city, and of its internal combustion (low emission) engine outside the city. "The use of the electric motor by itself," Sergio explains, "is not yet feasible, for three reasons: the cost, which is more than twice that of the car with an internal combustion engine; the limited range without the possibility of recharging; and finally—perhaps emphasized less but no less important—the speed. Finding yourself in the middle of highway traffic with a car that can't do more than 65 mph can make you panic, especially if you are being besieged and passed by big trucks. Panic is bad for people's health, and it's clear that people deserve ecological attention even before the environment does. The thermic engine is at the moment a great step forward, and we consider the hybrid Eta Beta the car of the moment, until the day when technology comes up with new and more salutary engines." Eta Beta is thus understood as a motory prosthesis, with adjustable components. So it is fitting that this MPV also has the capacity of growing longer or shorter, with a telelscopic maneuver, on the

CONFIGURAZIONE TARGA
CONFIGURAZIONE SPIDER
COFANI (ANT. + POST.) IN ALLUMINIO

FASE 1
FASE 2
FASE 3
SOSPENSIONE ANT.

Aluminum with its basic structure constitutes the backbone and the heart of the Argento Vivo and takes on a fundamental role in the play of the two-tone and the enlivening of the surfaces.

Wood and aluminum form the structure of the interior of the Argento Vivo prototype.

basis of the use demanded: for city driving, two plus two, short tail; for trips outside the city, four seats, long tail; for longer trips, with trunk, two seats and long tail. "The doors, in aluminum, are rectangular and very large, to facilitate passenger access. Complete with arched glass surfaces, they consist of a flat parallelogram, with a gull-wing movement, and pivot upwards with a complex movement expressly designed to obtain extremely reduced side dimensions. Without handles, the doors are operated entirely by remote control."

In the CNR Eta Beta research prototype technological devices are therefore necessary functions, not decoration, taken on equal terms with the noble materials that make up its architecture. Confirming the words of Augusto Morello: Mythos and Ethos set the pace in the humanizing of technologies in accelerated development and the automobile will thus signify more and more, in the intention of Pininfarina's design, seeming well, seeming opportune, with regard to man and for man. To that end, in 1997, "Pininfarina, with the Peugeot Nautilus, joins up with its extremely rich tradition in the field of style." So runs the official presentation. "A three-box four-seater sedan that rejects the grave, imposing stiffness of some models of this class, to espouse sleek and dynamic lines." If it's true that the three-box theme had become somewhat hackneyed in the configuration of the sedan, "We had already broken the mold," Sergio explains, "in 1958 with the Austin A40. And all the designers and constructors will break that mold with the MPV. That's why we thought we'd take a fresh look at the three-box, give it new dignity. And the time was just right, when the performance of the

airbag
rest
econ
25.4
auto
off
ON
Dolby system
FM 1 TON--GOLD

Three significant pictures of the CNR Eta Beta prototype in the urban context. Eta Beta represents a way of breaking with the concept of the automobile seen as a monolith.

sedans was closer all the time to the best-known sports cars."

Pininfarina studied and created the 2496 cc 6-cylinder Nautilus prototype over a Peugeot chassis, in honor of its active collaboration over half a century with the French car-maker. "The overall configuration recalls a fish shape," the official presentation continues. "A sensation due to the carbon-fiber spinal column that passes through the interior and holds the gear-lever turret and on-board controls, and to certain stylistic elements that recall the branchial cleft of a shark." The comparison with the shark arises, then, from a formal examination of the components that make up the definition of the prototype but, equally, from the harmonious behavior of the two architectures in motion—one natural, the other artificial. And here the artifice confirms the possibility that the equilibrium of all the technological components may work together for the definition of the harmonic identity of the car. In this sense the Nautilus "fits into the ample scenario of studies and research projects that Pininfarina, in its design center role, has always carried out in the automotive field, acting as a stimulus and a laboratory of ideas for the development of mobility... The Nautilus study privileges extremely vivacious forms, with soft features but sharp edges, in order to impart strength and character to the whole: the elongated hood gives a sense of power while the front, with its very low air intake, underlines the sporty nature of the car, which is also suggested by the design of the fenders, by the smallish size of the roof, by the elongated glass surfaces and by the reduced height of the roof panel... The Nautilus is a highly advanced proposal on the formal level, but it is still and above all a

CNR Eta Beta optimizes
the diversity of form
and of the surface treatment
of its components:
the aluminum of the doors,
of the roof, of the structures,
and the thermoplastic painted
resins all go to make up
our interpretation of the car.

functional prototype. The instruments, for example, can be read in two ways: digital for relaxed driving, and analogical for sports… The analogical instruments are installed on the dashboard in a sort of shell that opens up on command. The central spine contains two swivel-displays with which it is possible to control every function of the vehicle and to establish its exact position by means of a satellite navigation system. It represents the car's central nervous system and incorporates all its electronic equipment for interaction and vehicle control, as well as internet, teleconference and GPS connections. The interface was developed with the aim of reducing the presence of physical buttons to a minimum. Everything, wherever possible, has been virtualized. The body of the interface is an image on the displays, and interaction with this image (no longer a prerogative of the driver alone) is now possible through a unified system of control for all the occupants of the car. The third display, in the back, can disclose the information of the second (the first is constantly dedicated to the function of rearview mirror) and a small telecamera is equipped for teleconferences. Between the two back seats there is also accommodation for a portable PC that, in synergy with the informatics system of the vehicle, can effectively transform this part of the car into a fully-equipped, one-man mobile office." So, here, the hardware is matched by software designed to improve life on board. In addition to the instruments of safety control—that sort of virtual declamation of the diary of navigation—for the use of all the travelers, the apparatus also satisfies the needs of whoever, among them, cannot interrupt work begun in the office. The Peugeot Nautilus

CNR Eta Beta, 1996

Two pictures taken from Jules Verne's novel "20,000 Leagues under the Sea," showing the interior of the "Nautilus" submarine.

The Nautilus study privileges extremely vivacious forms, with soft features but sharp edges, imparting strength and character to the whole.

prototype is thus itself a prosthesis, where the performance goes far beyond the declaration of rapid mobility.

So with the Nautilus the three-space sedan is indeed revamped in form, corresponding however to technological advance in the broadest sense; in fact, in the Nautilus, state-of-the-art technology, including informatics, is part of the formal research of the prototype. This was not so with the Dardo, created in 1998 by Pininfarina on an Alfa Romeo 156 chassis; this car, on the contrary, instead of technological standardization aimed at excellence of form. "Today more than ever," the official presentation declares, "the automobile is not only a technological and functional product, but also and above all an object capable of arousing emotions: the design can play a role of great importance. It can really make the difference, it can be an added value that is decisive for the car's success… In fact it is the model's appeal, understood as the fusion of charm and aesthetic personality, that is the prime element of attraction for the user, the prime and decisive reason for buying the car—such competitive and fundamental requisites as quality, performance,

The Nautilus is a four-seater sedan with three volumes that rejects the grave, imposing stiffness of some models of this class, to espouse sleek and dynamic lines.

The Peugeot Nautilus is a highly advanced proposal on the formal level, but it is still and all a functional prototype.

The elongated hood gives a sense of power while the front, with its very low air intake, underlines the sporty nature of the car.

The instrument system can be read in two ways: digital reading for relaxed driving, and analogical for sports.

Alfa Romeo Dardo, 1998

The half-light accentuates the line of the Dardo prototype.

The form of the Alfa Romeo Dardo gives an impression of smoothness, of softness, but also of originality and vigor.

price and service being equal." If it's true that the automobile will become, more and more, a container of the most advanced and broadest technologies, this does not mean that technologies have to condition its formal equilibrium. This is the challenge of the Dardo. The Nautilus demonstrated the possibility of an aesthetic reconciliation of technics and aesthetics in the mobile object, setting itself up as a norm. But the Dardo goes further, affirming that design alone can make a qualitative difference in the quantitative standardization of technological invasiveness. And its sense is in the name: Dardo [Dart], launched onto the road of the future of the automobile by Pininfarina. "Deliberately detached from the wealth of past models, with no indulgence toward retro styling, the Dardo hurtles towards the future with aggressiveness,

Sergio, Lorenza, Paolo and Andrea Pininfarina with the Dardo prototype.

free from the classical legislative and industrial bonds to which a mass-produced car is subject. Dardo is the dream of a car destined to run in the wind. It is a complex form that in the dash of its lines and soft volumes shows off above all its pointed front, the three connecting lines between the shield and the large windshield, the front fender and the spoiler, the side panel and the rear wheel, and that recalls, in a spellbinding game, the design of the triangle, present on the front, sides and rear of the car, surmounted by the two headrests behind the driver and passenger seats... The form of the Dardo gives an impression of smoothness, of softness, but also of originality and vigor. It is a sensation that comes from many things, but that is grounded in particular upon what may be called the 'guidelines' of the Pininfarina study. The first guideline originates from the small shield, the Alfa Romeo logo, which frames and lends force to the imposing hood. It links up with the very short pillars of the windshield, thrusting the glass towards the sky. The second guideline goes in the same direction. It rises up from the front fender, just above the wheel, delimits the upper part of the body side and concludes in the rear wing. The third line moves in the opposite direction, another precise mark that from the rear fender cuts into the side panel, which contributes in turn to the triangular form obtained in it."

What shall we say about the sparkling of those ogives, cut out in the peak of the hood, that wink spent and dispense ironic flaming cones of light?... Dardo is not a functioning prototype, it is pure form. All the more necessary, when the complexity of form of the Nautilus is already simplification in the hands of the sorcerer's apprentices. But Dardo, as pure form, revives the sense of an ethics linked to the correspondence of intimate disposition, of ancient intention of design, of exquisite aesthetic nature, that wants the auto form and color, then content. Apprentices of all times and fields don't understand the difference, and it is to this that Dardo bears witness far more than the splendid Nautilus did. Dardo is thus in 1998 the function, well calculated by Pininfarina, that once again proves itself capable of giving form to seeming well, seeming opportune.

Two phases in the production of the research prototype.

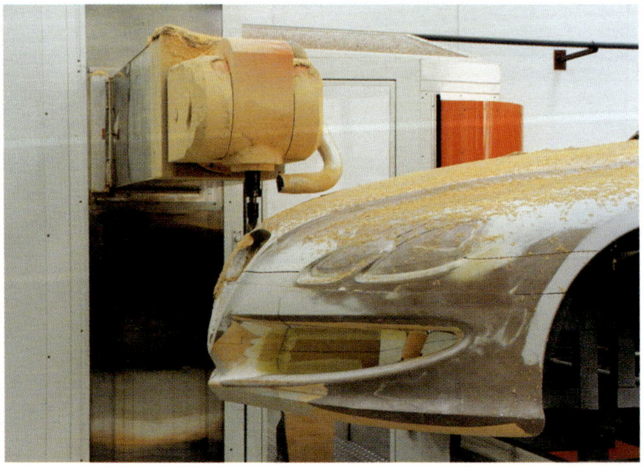

The Alfa Romeo Dardo hurtles towards the future with aggressiveness, free from the classical normative and industrial bonds to which a mass-produced car is subject.

Dardo is thus in 1998
the function, well calculated
by Pininfarina, that once again
proves itself capable of giving
form to seeming well, seeming
opportune.

Bibliography

Alvaro, C. "Fiori finti" [Artificial Flowers], in *Settantacinque racconti* [Seventy-five Stories], Bompiani, Milan 1955

Aspesi, N. *Il lusso & l'autarchia: storia dell'eleganza italiana (1930-1944)* [Luxury and Autarchy: A History of Italian Elegance], Rizzoli, Milan 1982

Atlante storico Garzanti: cronologia della storia universale, Garzanti, Milan 1994

L'auto italiana, November 1929, January 1930, June 1930

Balestra, N. *Intensamente Cisitalia*, document of the Associazione italiana per la storia dell'automobile, Museo dell'Automobile Carlo Biscaretti di Ruffia, Turin

Barthes, R. *Mythologies* [undated]

Battaglia, S. *Grande dizionario della lingua italiana*, Utet, Turin 1961

Baudrillard, J. *Le systeme des objets*, Gallimard, Paris 1968

Bellini, M. in *Modo*, nos. 76 & 77, 1985; *Domus*, no. 726, 1991

Bellucci, A. *L'automobile italiana (1918-1943)*, Laterza, Rome 1984

Bernabò, F., Rancati, G. in Alfieri, B. (ed.) *Catalogue Raisonné Pininfarina*, Automobilia, Milan 1990

Bessi, R. in Morandini, M. (ed.) *L'auto e il cinema*, Edizioni Alfa Romeo, Arese 1985

Biagi, E. *Il signor Fiat* [Mister Fiat], Rizzoli, Milan 1976

Boneschi, M. *Poveri ma belli: i nostri anni Cinquanta* [Poor but Beautiful: Our 1950s], Mondadori, Milan 1995

Bontempelli, M. "Da Roma..." [From Rome], in *Stato di grazia* [State of Grace], Sansoni, Florence 1942

Borgna, G. *Storia della canzone italiana* [History of Italian Song], Mondadori, Milan 1992

Bosoni, G., Nulli, A. "La storia del disegno industriale 1919-1990" in Castelnuovo, E. (ed.) *Il dominio del design 1919-1990*, vol.III, Electa, Milan 1991

Bossaglia, R. *Il Futurismo e l'automobile*, document of the Associazione italiana per la storia dell'automobile, Museo dell'Automobile Carlo Biscaretti di Ruffia, Turin 1998; *Il Novecento italiano* [The Italian Twentieth Century], Charta, Milan 1995

Branzi, A. (ed.) *Il disegno italiano 1964-1990*, Electa, Milan 1996

Burkhardt, F. *Marco Zanuso*, Federico Motta, Milan 1994

Buzzati, D. "Un amore" [A Love], in *Opere scelte*, Mondadori, Milan 1963

Caballo, E. *Pininfarina. Nato con l'automobile* [Born with the Automobile], Palazzi, Milan 1968

Calvino, I. "L'inseguimento" [Pursuit], in *Ti con zero* [t zero], Einaudi, Turin 1967

Carugati, D. in *Intervista* no. 14, July-August 1998; "Le donne i cavallier l'arme gli amori..." [Women cavaliers arms loves...] in *Stileindustria* no. 8, December 1996; *Capital*, February 1997; "La Porsche evoluzione design," in "Nautilus" (column), *Abitare* no. 379, December 1998

Casucci, P. *Jaguar, il fascino di un nome* [Jaguar, the Fascination of a Name], BL Italia, Rome 1979

Celant, G. *Auto Tattoo*, Automobilia, Milan 1986

Cerritelli, C. in Orsini, R. (ed.), *Dino Gavina*, Compositori, Bologna 1998

Ciferri, L. in Morello, A. *Sergio Pininfarina*, ADI, Milan 1985

Colombo, J. "Antidesign," in *Casabella* no. 342, 1969

Consolato Indiano [Indian Consolate]. *Documento*, Madhya Pradesh, 1988

Crispolti, E. *Omaggio a Luca Fontana*, Beniamino Carucci, Assisi-Rome 1971

D'Annunzio, G. *Forse che sì, forse che no* [Perhaps Yes, Perhaps No], Mondadori, Milan 1982

De Rosa, G. *Alfa Romeo Giulietta Spider*, Giorgio Nada, Milan 1994

Di Giammatteo, F. *Roberto Rossellini*, La Nuova Italia, Scandicci 1990

Enciclopedia della letteratura Garzanti, Garzanti, Milan 1997

Fagone, V. "L'arte all'ordine del giorno" [Art on the Agenda], in *Anni Trenta*, Mazzotta, Milan 1982

Felicioli, R.P. *Sergio Pininfarina. Car Man*, Automobilia, Milan 1998

Felicioli, R.P., Ramaciotti, L. *Pininfarina Ferrari 50 Figurini/Renderings/Esquisses*, Automobilia, Milan 1997

Fitzgerald, F.S. *Tender is the Night*, Penguin, 1939

Fitzgerald, F.S. *The Great Gatsby*, Penguin, 1926

Gadda, C.E. "L'uomo e la macchina" [Man and Machine], in *Le meraviglie d'Italia* [The Marvels of Italy], Einaudi, Turin 1964

Grandori, L. *Un sogno su quattro ruote* [A Dream on Four Wheels], Rizzoli, Milan 1990

Gregoretti, U. (ed.) *Fonografo italiano. Raccolta di vecchie incisioni 1890-1940* [Italian Phonograph. A Collection of Old Recordings], Nuova Fonit Cetra, II serie, Milan [undated]

Jansen, P.G. in *La Sera*, 14 April 1939
Joyce, J. "After the Race," in *Dubliners*, from *The Essential James Joyce*, Granada, London 1977
Leuschel, K.S., Oelke, H. *Design. Sviluppi in Germania* [Design. Developments in Germany], Internationes, Bonn 1992
Madaro, G. *Alfa Romeo Duetto*, Giorgio Nada, Milan 1990
Marinetti, F.T. "Manifesto del Futurismo" in *Figaro*, Paris, 20 February 1909; "La nuova religione morale della velocità" [The New Moral Religion of Speed], in *L'Italia futurista* no. 1, Milan, 11 May 1916
Mereghetti, P. (ed.) *Dizionario dei film*, Baldini & Castoldi, Milan 1997
Morandini, M. (ed.) *L'auto e il cinema*, Edizioni Alfa Romeo, Arese 1985; *Sessappiglio: gli anni d'oro del teatro di rivista* [Sessappiglio: The Golden Years of Theatrical Revues], Il formichiere, Milan 1978
Moravia, A. *The Time of Indifference* [Gli indifferenti], trans. Angus Davidson, Penguin, 1970; "Scherzi di Ferragosto" [August Holiday Pranks], in *Racconti romani* [Roman Tales], Bompiani, Milan 1954
Morello, A. *Sergio Pininfarina,* ADI, Milan 1995
La nuova enciclopedia dell'arte Garzanti, Garzanti, Milan 1986
Palazzeschi, A. *Sorelle Materassi* [The Materassi Sisters], Vallecchi, Florence 1970
Pansera, A. "Protodesign italiano," in *Anni Trenta*, Mazzotta, Milan 1982
Pininfarina, Sessant'anni, Edizioni Pininfarina, Turin 1990
Piovene, G. *Viaggio in Italia*, Mondadori, Milan 1957
Pirlo, V. *D'Annunzio e l'automobile*, document of the Associazione italiana per la storia dell'automobile, Museo dell'Automobile Carlo Biscaretti di Ruffia, Turin 1997
Proverbio, P. in *Abitare* no. 380, January 1999
Rizzo, S. (ed. and trans.) *Blues, ballate e canzoni. Bob Dylan*, introduction by Pivano, F., Newton Compton, Rome 1972
Ramaciotti, L. *Solitaires*, Automobilia, Milan 1989
Ramaciotti, L., Alfieri, B. *Mythos*, Automobilia, Milan 1989
Rogliatti, G. *Ferrari & Pininfarina*, special edition of *Ferrari Story*, Turin 1987
Rumi, G.(ed.) *Vita civile degli italiani. Società economia cultura materiale. Trasformazioni economiche mutamenti sociali e nuovi miti collettivi 1920-1960* [The Civil Life of the Italians. Society Economy Culture Material. Economic Transformations Social Changes and New Collective Myths], VI vol., Electa, Milan 1991
Santini, A. *Nuvolari. Il mantovano volante* [Nuvolari. The Flying Mantuan], Rizzoli, Milan 1983
Segoni, R. *La carrozzeria Pininfarina vista da...* [Pininfarina Bodywork Seen By...], document of the Associazione italiana per la storia dell'automobile, Museo dell'Automobile Carlo Biscaretti di Ruffia, Turin 1997
Tambini, M. *Il look del secolo* [The Look of the Century], Mondadori, Milan 1997
Thomas, P.-Y. "Armonia innovatrice" [Innovative Harmony] in Alfieri, B.(ed.) *Catalogue Raisonné Pininfarina*, Automobilia, Milan 1990
Tonelli Michail, M.C. *Il design in Italia 1925-1943*, Laterza, Rome 1987
Treccani, *Dizionario Enciclopedico*
Webb, C. *The Graduate,* [undated]

Photographic Credits

The photographs published in this volume have been kindly supplied by the Archivio Storico Pininfarina, Turin, which has authorized their reproduction:
Archives: Alfa Romeo, Fiat, Ferrari, Lancia, Peugeot; Aufnahme, Julius Weitmann, Stuttgart; Bertazzini, Turin; Giorgio Bellia, Turin; Jo Bilbao; Foto Torazza, Turin; Foto Terreni; Giancolombo, Milan; Enzo Isaia, Turin; Cedomir V. Komljenovic; P. Lecrêque, Paris; Rodolfo Mailander, Rome; Alberto Martinez, Paris; Moisio, Turin; Moncalvo fotografie, Turin; Stefano Mu, Turin; Antonia Mulas © 1999, Milan; Ugo Mulas © 1999, Milan; Renzo Muratori, Turin; Remo Pecorara, Turin; Publifoto, Turin; Foto RG, Turin; Foto Valtorta, Montecatini Terme.

Photographs have been also supplied by:
AFE, Archivio Storico del Cinema, Rome
Archivio E. Castruccio, Milan
Farabolafoto, Milan

This volume was printed by Elemond S.p.A.
at the plant in Martellago (Venice) in the year 1999